DISEASES AND DISORDERS OF FOREST TREES

A GUIDE TO IDENTIFYING CAUSES OF ILL-HEALTH IN WOODS AND PLANTATIONS

by S C Gregory and D B Redfern
Forestry Commission

London:

Front cover, top left: *Peridermium pini* on Scots pine; pustules on a branch canker (see Figure 42 in main text).

Front cover, top centre: *Ramichloridium pini* on lodgepole pine (see Figure 45 in main text).

Front cover, top right: Oak leaves with spots and blotches caused by *Apiognomonia errabunda* (see Figure 60 in main text).

Front cover, bottom left: Scots pine plantation in which most of the trees are dead or dying as a result of *Brunchorstia* infection (see Figure 46 in main text).

Front cover, bottom centre: Fruit bodies of *Lachnellula willkommii* (see Figure 52 in main text).

Front cover, bottom right: Atmospheric pollution injury in Norway spruce (see Figure 5 in main text).

Back cover, left: Killing of shoot tip and buds of Sitka spruce by autumn frost (see Figure 30 in main text).

Back cover, centre: Beech bark disease; "tarry spots" caused by *Nectria* infection (see Figure 63 in main text).

Back cover, right: Breakage of a larch stem caused by *Phaeolus schweinitzii* decay (see Figure 71 in main text).

CONTENTS

CONTENTS

INTRODUCTION

A large number of diseases and disorders can affect forest trees in Britain. Confidence in identifying particular causes from this range of potential problems requires a combination of knowledge and experience that is not easily acquired without becoming a specialist in the field. Nevertheless, we hope that part I of this handbook will provide some useful guidelines by which managers and owners can investigate health problems in their trees. Straightforward field observations of the kind covered here can often allow crucial distinctions to be made between broad categories of disease; they can also provide solutions on which practical decisions can be based.

Part II of the handbook is a key to the most common problems of plantation trees and in part III important features of some of the diseases and disorders covered in part II are summarized. Full details of these can be found in "Diseases of Forest and Ornamental Trees" by D H Phillips and D A Burdekin (Macmillan; 2nd ed; 1992), which also provides descriptions of the problems that are noted in the key but not treated in part III.

The key and descriptions include some common diseases of trees that are not normally regarded as major crop trees but which are frequently grown in commercial woodlands. However, street and garden trees are not dealt with specifically. Although these may suffer from some of the same diseases and disorders as plantation trees, symptom expression and pattern are often different, reflecting the different situations in which they are grown. For a comprehensive treatment of the problems of ornamental and amenity trees, we refer readers to "Diagnosis of Ill-health in Trees" by R G Strouts and T G Winter (HMSO; 1994).

We have not dealt separately with Christmas tree plantations since, by and large, they are likely to suffer from the same diseases and disorders as forestry plantations of the same age. Methods used in the cultivation of Christmas trees probably expose them to higher risks of some forms of injury – herbicide damage, for example – and it should be acknowledged that some problems that are significant to Christmas trees but of trivial importance to forest crops have not yet been fully investigated. A common example is summer needle browning of *Abies* species, the cause of which is unknown.

It is also relevant to note here that methods of chemical control have not generally been developed or approved for diseases of forest trees, as very few diseases would ever justify such treatment on a forest scale in Britain. With the single current exception of stump treatment against *Heterobasidion annosum* (Fomes), forest diseases are either tolerated or managed; sometimes they can be avoided altogether by the choice of species or silvicultural technique.

We have made several references to damage by insects. While many insect attacks are made quite obvious by the type of damage or the presence of the culprits themselves, other cases are not clearly insect-related and so could not readily be excluded from a general diagnostic guide. We are indebted to Mr T G Winter and Mr C I Carter for expert assistance with our brief coverage of insects. Further information can be found in "Forest Insects" by D Bevan (Forestry Commission Handbook 1; HMSO; 1987), and in "Christmas Tree Pests" by Clive Carter and Tim Winter (Forestry Commission Field Book 17; The Stationery Office; 1997).

Finally, we should emphasize that we have concentrated on the commonest and most important damaging agents in woods and plantations. There are some uncommon diseases and disorders which can produce similar effects to those we describe. Consequently, not all forms of damage will be identified by the key. If, on investigation, a problem looks serious, it may be advisable to seek confirmation of its cause from a specialist.

NAMES

The tree species for which we have used common names in the text are listed opposite with their botanical (Latin) names. The naming of pathogens and the diseases they cause has proved less straightforward as many do not have common names in Britain. Rather than contrive names in English, we have used the Latin name of the pathogen – as in *Lophodermium* needle-cast and *Crumenulopsis* canker. However, although the use of Latin names is supposed to avoid the confusion and imprecision of common names, it frequently fails on the first count because the very accuracy of the system entails frequent revision of names by researchers. In consequence, several fungi have one or more aliases from the inheritance of older names. Where they might be useful for referring to older books, we have given some of these earlier variants in addition to the currently accepted name.

Corsican pine	*Pinus nigra* var. *maritima* (Aiton) Melville
Douglas fir	*Pseudotsuga menziesii* (Mirbel) Franco
European larch	*Larix decidua* Miller
grand fir	*Abies grandis* (Douglas) Lindley
hybrid larch	*Larix* x *eurolepis* Henry
Leyland cypress	x *Cupressocyparis leylandii* (Jackson & Dallimore) Dallimore
lodgepole pine	*Pinus contorta* Douglas
noble fir	*Abies procera* Rehder
Norway spruce	*Picea abies* (L.) Karsten
Sitka spruce	*Picea sitchensis* (Bongard) Carrière
Scots pine	*Pinus sylvestris* L.
ash	*Fraxinus excelsior* L.
aspen	*Populus tremula* L.
beech	*Fagus sylvatica* L.
common alder	*Alnus glutinosa* (L.) Gaertner
crack willow	*Salix fragilis* L.
cricket bat willow	*Salix alba* var. *caerulea* Smith
gean	*Prunus avium* L.
hawthorn	*Crataegus monogyna* Jacquin
hornbeam	*Carpinus betulus* L.
horse chestnut	*Aesculus hippocastanum* L.
plum	*Prunus domestica* L.
rowan	*Sorbus aucuparia* L.
sycamore	*Acer pseudoplatanus* L.
wych elm	*Ulmus glabra* Hudson

PART I

A GUIDE TO DIAGNOSING TREE PROBLEMS

1. GENERAL CONSIDERATIONS

Although there are occasions when background information – such as site history and meteorological records – offers vital clues or valuable supporting evidence, most information relevant to making a diagnosis has to be drawn from examination of the affected trees. The most valuable field observations are generally those relating to the **symptoms**, **distribution** and **timing** of the damage and the following sections are concerned with some of the practical aspects of collecting this information. There are so many situations in which tree damage can occur that it would be impractical to attempt to formulate rules of procedure to cover every case, and anyone who is regularly called on to look into tree health problems will develop their own method of working. However, our experience of diagnostic and advisory work suggests a number of principles that are worth keeping in mind during any investigation.

i. Many forms of damage tend to look the same from a distance. It is advisable to take a closer look at damaged trees even when the probable cause is known. There is no disadvantage in becoming more familiar with a common problem – and always a chance of turning up something unexpected.

ii. Second-hand information should be treated with scepticism, and you should also be suspicious of your own unsupported recollections. It is advisable to write down observations as soon as possible after making them.

iii. Non-living agents – weather, herbicides, site conditions – are at least as likely to be involved as living agents – fungi, bacteria, insects.

iv. A mundane explanation, like frost or herbicide misuse, is far more likely than a bizarre or sinister one such as radioactivity, jettisoned aircraft fuel, or a disease new to Britain.

v. Although they are not common events, new pests and pathogens, and new forms of behaviour in well-known ones, do arise. As they may then have very serious consequences, these possibilities should not be overlooked.

vi. What appears to be a single problem may in fact be two or more quite separate forms of damage. Especially thorough investigation is necessary when trees of several species over a wide area seem (or are alleged) to have suffered the same type of damage.

vii. Some pathogens are primary and will attack previously healthy trees. Others are secondary and only cause disease in trees which have been injured or weakened. Some common pathogens, such as *Armillaria*, can act in both capacities and so require extra care in diagnosis.

A final, though extremely important, general point is that a sound knowledge of the anatomy and physiology of the trees to be examined is the foundation for accurate assessments of condition, and for successful diagnosis. This can hardly be overemphasized: a significant number of the enquiries dealt with by the advisory service of the Forestry Commission Research Agency are eventually resolved as "non-problems". These reflect the frequency with which a sinister interpretation is put on features that are unusual but still within the normal, healthy range of variation of the species in question. For example, it is quite usual for some pines to retain needles for only two years and when the older needles change colour in autumn, plantations can assume a rather alarming yellow appearance.

2. FIELD INVESTIGATION

2.1. Equipment and Sampling

The following basic equipment is worth having to hand on any field visit, even if damage investigation is not planned:

- binoculars;
- notebook;
- something to dig with (a trowel is often good enough – especially if it has to be carried any distance);
- a sharp, strong knife;
- a tool, such as a small axe or chisel, capable of cutting into thick mainstem bark;
- secateurs;
- polythene sample bags of various sizes.

If samples are collected for laboratory examination, they should represent a range of symptoms and, most importantly, include examples of the earliest stages of damage. In the case of shoots, samples should have, whenever possible, intact junctions between dead and live tissue. If a pathogen is responsible, it is most likely to be active (and hence identifiable) there. Although it is generally advisable to store samples in polythene bags, soft fungal fruit bodies are an exception. They rot very quickly in polythene and are better kept in paper, even if they start to dry out.

2.2. Symptoms and their Assessment

2.2.1. *Foliage Discoloration and Die-back*

Accurate assessment of symptoms requires close examination. An apparently uniform yellowing seen from some distance away might, in fact, be the result of discoloration only in old foliage, or only in young foliage; or it might represent leaves with yellow blotches, or entirely yellow leaves. Each pattern would suggest a different range of possible causes. Equally demanding of careful scrutiny is the possibility that the cause of damage is physically separate from the most noticeable symptoms. Thus a branch with yellow or brown foliage could be suffering the effect of a girdling injury or bark infection some distance back along the branch system.

In cases where foliage is discoloured, it is important to know whether the bark and buds of affected shoots are alive or dead. The distribution of dead and live tissues can be determined by cutting into them with a sharp knife. Beneath the brown outer bark, live bark is firm, moist and light-coloured (pale pink, pale green or white). Inside the bud scales, live buds are similar in texture and colour. Dead or dying tissues are light or dark brown, or black; they are often dry and shrunken but sometimes water-soaked or soggy. Areas of dead bark on vigorously growing stems and shoots are often noticeable as depressions. If shoots are dead, it is advisable to locate the live tissues on the older parts of several shoot systems in order to establish whether there is any pattern to the age of tissues that are damaged.

The pattern and distribution of damaged shoots and foliage in the crown, and the distribution of damaged foliage over affected shoots, can be highly significant. Some shoot-disease fungi typically cause damage in the lower crown and several foliage diseases of conifers only affect needles of a particular year. Wind-related phenomena – exposure, spray drift and some foliage diseases, for example – often occur only on one side of the crown or branch (figs 2 and 6). When the whole crown of a tree appears to be damaged, it is important to check for surviving shoots. The presence of only a few of these, shoots protected by weed wound-infecting fungi or root-disease fungi that spread up from the roots to cause butt-rot. It is worth remembering that drought cracks also function as wounds, at least while they are open, and can lead to stain even though they might eventually become buried inside the stem and barely visible as a physical defect.

2.3. Distribution of Damage

There are many cases in which the extent and distribution of damage in the crop are as important to diagnosis as the nature of the symptoms. Even if only a small group or a single tree appears to be affected at first glance, a wider look around the stand is worthwhile, especially if a major crop species is involved. Extensive inspection may be required in serious cases and, when there is high mortality, persistent searching is sometimes necessary to find survivors with lesser symptoms that might offer some clues about the cause. Normally, the confidence with which a final diagnosis can be made improves as observations become more comprehensive.

The key questions about distribution include the following:

are all trees of the same species or age class affected?

if not, are damaged trees in groups, or scattered?

is more than one species or age class affected?

is damage confined to a particular area?

does the distribution of damage bear any relationship to topographic or artificial features such as hollows, road-sides or stand edges?

Although some forms of damage characteristically occur in particular patterns, these are not immutable. Injury by radiation frosts is certainly more common in "frost hollows" but it is by no means confined to them. Several factors, such as nutritional status, can upset this commonly accepted association. Root diseases like *Armillaria* and *Heterobasidion* that are causes of "group killing" frequently kill scattered individuals as well. Moreover, the identification of group killing itself requires some care as a grouped or patchy pattern of damage can be imposed on many potentially widespread agents by accident, topography or soil type.

2.4. Timing the Occurrence of Damage

Concurrently with the investigation of what has happened and where, evidence of when it happened should be sought. On partially damaged trees, the disturbance to the normal growth pattern can give quite an accurate idea of when damage occurred, especially in trees such as conifers that have a regular and predictable growth habit. In most trees from temperate parts of the world, the part of a shoot between successive terminal bud scars (if they can be recognized) represents a year's growth. In many conifers, the year is further marked by the annual production of a whorl of side branches so that dating die-back can be relatively easy (fig 1). In cross-sections of stems, abnormal patterns in the production of annual growth rings can, in many cases, indicate when damage occurred, but interpretation of these features usually requires laboratory examination.

Fig 1. Dating the occurrence of damage. The shaded shoot was formed in 1994; the presence of a fully developed but unflushed bud indicates that death occurred between the end of the 1994 growing season and the beginning of the 1995 growing season.

Whether external or internal evidence of the date of injury is used, reliable results can only be achieved with material that provides continuity between live and dead tissue such that the former gives the current year as a baseline. Consequently, samples for dating have to be carefully chosen and, even then, there are pitfalls in interpreting shoot growth or annual ring patterns. Apparently regular growth patterns can be complicated by, for example, false rings, lammas growth, formation of internodal shoots, development of adventitious buds, and forms of damage (such as browsing) that remove shoots completely.

The state of the damaged tissue itself can indicate when damage occurred. Shoots that are killed early in the growing season, before they are fully hardened, tend to hook over, and may lack terminal buds. At the extreme, newly flushed shoots will collapse completely, but may nevertheless persist to provide evidence. On broadleaves, leaves are often produced successively so that damage confined to fully expanded leaves at the base of the current shoot would suggest an injury that occurred early in the season and did not recur.

In the case of completely dead trees, or branch systems, it is sometimes possible to make qualitative judgements about when death occurred from the state of deterioration. Dead conifers tend to retain foliage only for a year or so, and an intact fine twig structure for perhaps another year or two. However, these times can vary tremendously with site, species and climate. Main stem bark starts to detach from about three to five years after death, though this can be much more rapid if bark-boring insects attack the dying tree.

If there are clues to when damage occurred, it is important to establish whether it all took place at the same time or appeared to spread and develop over a period. This can allow a distinction between one-off events, such as many climatic and chemical injuries, and slowly spreading problems, such as fungal root diseases. Nevertheless, it is as well to be wary of the apparently straightforward. The effects of a single incident may develop slowly and at different rates in different individuals – there have been cases of April frost injury in spruces where full symptom expression did not occur in some trees until the following dormant season. By contrast, a long process of infection and spread by a fungus inside a tree may culminate in apparently sudden browning. Since the life cycles and activity of many pathogens are regulated by season and climate, infection and symptom expression may be simultaneous over large areas, giving an impression of "scorching" induced by the direct effects of climatic or chemical injury.

3. DIAGNOSING THE CAUSE OF DAMAGE

As a picture is built up of symptoms, distribution and timing, the key in the next section can be attempted. This will lead to broad categories of solution ("climatic injury", for example) in some cases and more precise explanations in others – usually where more information is available. Even then it may be worthwhile to seek supplementary or confirmatory evidence by means of further investigation. With problems caused by humans this can be as simple as examining sites that did not receive the suspect treatment. However, it should be emphasized that, with some diseases, a completely confident diagnosis cannot be made without laboratory investigation.

PART II

KEY TO MAJOR DISEASES AND DISORDERS OF PLANTATIONS AND WOODLANDS

The development of diseases and disorders in forest trees is extremely variable and the end results – foliage discoloration and die-back – are similar in many cases. Consequently, it is not practicable to devise an identification key that presents completely unequivocal choices or pinpoints individual agents with absolute certainty. The following key is designed to provide the most likely cause, or group of causes, for the observed damage, and to indicate when further advice should be sought. There are many rare diseases and disorders that are not dealt with, though some that are potentially serious and that might be confused with more prosaic problems are indicated. The more common and important problems are more fully described in part III; further details of the pests, diseases and disorders that are not dealt with there can be found in "Diagnosis of Ill-health in Trees" (R G Strouts and T G Winter; HMSO; 1994), "Diseases of Forest and Ornamental Trees" (D H Phillips and D A Burdekin; Macmillan; 2nd ed.; 1992), or "Forest Insects" (D Bevan; Forestry Commission Handbook 1; HMSO; 1987).

For those not familiar with identification keys, each set of lettered paragraphs with the same number (e.g. 1a–1c) gives a series of alternatives. Read them all, choose the one that best fits the situation, and it will give another number to look for, or suggest a likely cause, or suggest seeking expert advice. The last is the best option if none of the alternatives fit. If two or more seem to fit, follow each pathway in turn as there may be more than one problem involved.

In the key, we have used "foliage" to refer to both conifers and broadleaves, but "leaves" or "needles" to refer to broadleaves or conifers individually.

THE KEY

1a Discoloration, die-back and death of trees within a year of planting **2**

1b Damaged trees not in first year after planting **3**

2a Stem girdled by bark removal or tunnelling, or roots debarked, tunnelled or bitten off; this must be investigated on carefully lifted plants

- **insects** or **voles**; see part III, section 7.1.

2b Not so

- likely to be either **planting failure** or **herbicide injury** (part III, section 1.1), especially if a high proportion of plants is affected. However, in Corsican pine, the possibilities of ***Brunchorstia*** disease (part III, section 4.6.4) and **spring frost** damage (part III, section 4.6.7) should be considered; in all species **drought** is a further possibility in exceptionally dry seasons, though the effects of drought are often exaggerated.

3a Crown symptoms associated with girdling of branches or main stem by physical damage (tunnelling, mining or bark removal)

- **insects** or **mammals**. See part III, section 7.1. This kind of damage is surprisingly easy to overlook and it should be ruled out before going further, especially in the following cases: **small trees with their bases obscured by vegetation; crown die-back in sycamore or beech; large conifers with resin bleeding or resin tubes on the main stem.** Note that the last could be due to attack by bark beetles. Such attacks are generally stress-related but could involve dangerous non-indigenous species; advice should be sought from the Forestry Commission where this is suspected.

3b No physical damage **4**

4a Browning, die-back and mortality of pole-stage or older Norway spruce along stand edges but with healthy trees mixed among affected trees; roots still alive in trees with advanced die-back

- likely to be **top-dying**; see part III, section 3.4.

4b In thicket to pole-stage lodgepole pine; browning, defoliation and die-back of many trees; affected trees having most or all needles, including current needles, removed or reduced to ragged stumps; defoliation starting early in the season on extending shoots (which may collapse) and moving down to older needles

- pine beauty moth *(Panolis flammea)*.

Key Section continues

4c In thicket stage or older Sitka spruce; loss of all but the current year's needles in many trees over all or part of the crown, but especially in the lower crown

- likely to be ***Elatobium***; see part III, section 3.2.

4d Not fitting any of these descriptions **5**

5a Browning and/or die-back in many trees making a clear "edge effect"; all or nearly all trees closest to the plantation edge affected (though not necessarily to the same degree) but the severity and prevalence of damage decreasing away from the edge; on individual trees, damage may occur only on, or be worse on, the side closest to the plantation edge **6**

5b Not so **7**

6a Crop over 10 years old; Norway spruce only affected

- **climate**, an **airborne toxin** or **top-dying**. Airborne toxin is a highly unlikely explanation if Norway spruce is the only species affected in a mixed crop. See part III, sections 1.2, 1.3 and 3.4.

6b Crop over 10 years old; Scots pine only affected

- **climate**, an **airborne toxin** or ***Lophodermium*** **needle-cast**. Airborne toxin is a highly unlikely explanation if Scots pine is the only species affected in a mixed crop. See part III, sections 1.2 and 4.3.

6c Not as a or b

- **climate** or an **airborne toxin** (including **road-salt spray**); see part III, sections 1.2 and 1.3.

7a Damage in a young crop that has recently been chemically weeded; the distribution of damage corresponding with the pattern of weed control

- likely to be **herbicide damage**. Laboratory analysis might be necessary to confirm this. See part III, section 1.3.1.

7b Damage in a young crop that has recently been fertilized; developing more or less simultaneously in several, possibly scattered, trees; browning, shoot death and mortality; damage to ground vegetation associated with some of the more severely damaged trees

- likely to be **fertilizer damage** but laboratory analysis might be necessary to confirm this. See part III, section 1.3.1.

7c Fitting neither description **8**

8a **Yellowing, stunting, and possibly some die-back of young Sitka or Norway spruce growing in heather; affecting many trees uniformly, often in patches**

- likely to be **"heather check"** (severe nitrogen deficiency); see part III, section 3.5.

8b **Site on shallow soil with high chalk or limestone content; widespread yellowing and decline of crop or large patches of it; progressing to loss of apical dominance (in larch), die-back and mortality**

- likely to be **lime-induced chlorosis**; see part III, section 1.5.3.

Note In a and b the most likely explanation based on site type is given. Nevertheless, some caution is necessary over the possibilities of ***Elatobium*** damage in spruces (part III, section 3.2), ***Lophodermium*** **needle-cast** and related problems in Scots pine (part III, sections 4.3 and 4.6.5), and, in all conifers, **root diseases** (part III, section 2). The last can be severe on high pH sites.

8c **Site/symptoms fitting neither description 9**

9a **Several genera in the same area show similar symptoms of foliage discoloration and die-back though not necessarily to the same degree 10**

9b **Damage in a monoculture or only one species affected ... 12**

10a **Damage occurs in a crop with one or more of the following characteristics:**
- recently treated with herbicide or fertilizer
- adjacent to an area so treated
- within 1–2 km of an industrial plant
- adjacent to a major road
- in an exposed situation within 5 km of the coast
- planted on a reclaimed industrial site

- **chemical injury** is a strong possibility but specialist advice might be necessary to confirm this or rule it out. See part III, sections 1.2, 1.3 and 1.5.2.

10b **Not in any of these situations 11**

11a **Damage occurring in the dormant season or spring; foliage browning and/or die-back of 1–2 year shoots of conifers, or death of shoot tips and shrivelling of new growth in conifers and broadleaves**

- likely to be **frost** or **winter injury** but these can be difficult to diagnose with certainty. If damage occurs on a large scale, it might warrant investigation by a specialist. See part III, sections 1.2, 1.4.1. and 1.4.2.

Key Section continues

11b **Not fitting this description**

- likely to be a site-related problem or coincident occurrences of several diseases and disorders; attack by a disease with a broad host range is a less likely possibility – go to **12** and proceed separately for each species affected.

12a **Damage confined to foliage; the overwhelming majority of shoots and buds remain alive** **13**

12b **Damage involves other parts of the tree(s); foliage symptoms may be absent, or consequent on root/shoot damage, or precede it** **52**

13a **Foliage discoloration confined to a single branch; adjacent branches healthy**

- most likely to arise from the effect of a **bark disease**; examine the affected branch carefully and if signs of a lesion or canker can be found, go to **52**. Less likely possibilities are the early stages of a **wilt disease** (e.g., Dutch elm disease and watermark disease of willow – part III, sections 6.9 and 6.11) or a **chimera** (part III, section 1.6).

13b **Damaged foliage variously distributed through the crown but not confined to a single branch** **14**

14a **Damage obviously due to physical removal or abrasion of tissue, consisting of holes, ragged edges, mines, or cavities** **15**

14b **Leaves of oak, sycamore or hawthorn patchily or completely discoloured (when they may be slightly curled or puckered) and covered in white powdery deposit**

- mildew.

14c **Not so** **16**

15a **On *Prunus*; damage consisting of more or less circular holes, 2–3 mm in diameter**

- shothole phase of **bacterial canker** or ***Blumeriella*** disease; see part III, section 6.6.1.

15b **Not so**

- not further keyed; likely to be **wind abrasion** or **insect damage**.

16a Leaves/needles normal size but deformed by outgrowths (pimples, blisters, etc) or by puckering, twisting, or rolling **44**

16b Leaves/needles abnormally small **51**

16c Not with these symptoms; leaves/needles may be withered but are otherwise of more or less normal size and shape, and are intact **17**

17a Affected foliage has prominent, white, yellow, orange or brown powdery pustules on discrete spots, blotches or (on needles) bands **32**

17b Affected foliage without pustules but having discrete random spots, blotches or (on needles) bands, most of which are away from the edge (of leaves) or tip **33**

17c Affected foliage discoloured in a regular pattern between veins, or at tip (of leaves or needles), or around margins; damage may be over whole crown or on one side **43**

17d Not with any of these distinctive patterns **18**

18a Foliage with a general pale green or yellow–green tinge without a clear pattern; colour may be concentrated in certain areas of individual leaves or needles but is not confined to discrete spots or blotches; widespread in crop or in large patches; may be associated with poor growth or form

- probably a **nutritional disorder** or other **site-related problem**; see part III, section 1.5.

18b Foliage markedly discoloured yellow, brown, grey or black (may also be withered or absent); affected leaves or needles completely discoloured or with only small portions staying green **19**

19a On conifers; predominant colour yellow; bright or dull yellow, or almost white **20**

19b Predominant discoloration brown, grey, or black, or a mixture of colours, including yellow (may also be withered or absent) **21**

20a Affecting single conifers (rarely two or three in a stand); needles uniform, brilliant lemon yellow or almost white; may affect whole or part of tree, or single shoots or all needles on one side of shoot

- **chimera**; see part III, section 1.6.

Key Section continues

20b On lodgepole pine; yellow (and brown) needles in bunches at the tips of scattered shoots

- probably ***Ramichloridium***; see part III, section 4.6.3.

20c On pine or spruce; dull or dirty yellow discoloration of many or all older needles; affected trees on peaty ground; discoloration may particularly affect the lower or mid crown in larger trees; in spruce current needles remain green; in pines they may have yellow tips

- probably **potassium deficiency**; see part III, sections 3.5 and 4.1.

20d On spruce; yellow-green colour affecting all foliage or most marked in youngest needles; may be accompanied by growth reduction

- probably **nitrogen deficiency**; see part III, section 3.5.

21a On conifers **22**

21b On broadleaves **31**

22a On larch

- likely to be **frost** or ***Meria laricis*** (part III, section 5.1); a less likely possibility is the insect ***Coleophora laricella*** which mines needles and causes them to wither.

22b On Douglas or grand fir; browning in late winter to spring on scattered shoots of the previous year; may be accompanied by die-back and reduced extension growth in spring

- probably **winter injury**; see part III, section 1.4.2.

22c On young noble fir; current needles develop reddish brown bands or become completely brown, or possibly withered, during summer; may affect few or many needles variously distributed on shoots

- **summer needle browning, cause unknown**, a not infrequent problem in Christmas tree crops.

22d On spruce **23**

22e On pine **27**

23a On Norway spruce over 10 years old; foliage browning from tips of shoots backwards (fig 31); whole crown may be affected, or patches, or windward side; usually affecting several trees in a stand

- early stages of **top-dying**; see part III, section 3.4.

23b Not so **24**

24a In summer to autumn on trees up to about 10 years; discoloration associated with light silky webbing; progressing from centre of trees outwards; needles at first speckled (<u>not</u> blotched) dull yellow then entirely bronze or coppery brown; current foliage is affected last

- conifer spinning mite.

24b Not so **25**

25a In summer to autumn; previous year's needles blotchy yellow or brown (or detached); needles at base of current shoot yellow or spotted yellow or absent; small current shoots on semi-suppressed branches in the lower crown may be completely straw-coloured; sticky and sooty deposits may be evident on shoots and foliage

- ***Elatobium***; see part III, section 3.2. Note that this is common on Sitka spruce, less so on Norway spruce.

25b In autumn to winter; only youngest needles affected, uniform red–brown (though the underside may be pinkish); needles at base or tip of shoot (around bud) may remain green; whole crown of trees up to 2 m affected, or lower branches on larger trees

- **autumn frost**; see part III, section 3.1. Note that this is common on Sitka spruce; rare on Norway spruce.

25c Discoloration in neither of these patterns **26**

26a In autumn or winter; severe discoloration of virtually all foliage associated with sooty and sticky deposits on shoots and foliage; on trees up to about 10 years old

- likely to be the result of an autumn/winter attack by ***Elatobium***; see part III, section 3.2.

26b Severe discoloration of virtually all foliage on trees up to about 10 years old but no sooty or sticky deposits; many trees affected

- possibly attack by mites (**conifer spinning mite** or ***Nalepella*** spp.) or some form of **abiotic injury**. However, as the sooty and sticky deposits normally associated with ***Elatobium*** may be washed off by rain, this too cannot be ruled out. Laboratory examination may be required for diagnosis.

27a On Scots pine **28**

27b On Corsican pine **29**

27c On lodgepole pine **30**

28a **In late winter to spring; wholesale browning of foliage over whole crown or on windward side, or noticeably confined to lower crown; occurring in crops of all ages, often affecting whole plantations or groups of Scots pine in a matrix of other species; symptoms remaining confined to older needles as trees flush**

- probably ***Lophodermium* needle-cast** (see part III, section 4.3) but possibly **winter injury** (section 4.6.2.).

28b **Discoloration of current needles in late summer; a small proportion of individuals may be severely discoloured with the remainder being unaffected**

- probably ***Lophodermella***; see part III, section 4.4.

29a **On trees <1 m; red–brown discoloration in spring/early summer**

- probably **spring frost injury**; see part III, section 4.6.7.

29b **On trees of any age; discoloration developing on current foliage from late summer to autumn; greyish brown by following spring; bases of affected needles may remain green**

- probably ***Lophodermella***; see part III, section 4.4.

30a **In autumn to winter; yellow and brown needles at the very tip of the current shoot; affecting scattered shoots in the lower to mid-crown**

- probably the early stages of ***Ramichloridium* die-back**; see part III, section 4.6.3.

30b **In late winter to early summer; discoloration of previous year's needles along most of shoot length (i.e. not just at tip); progressing from olive-green to pale brown and then red-brown; affecting scattered shoots; may occur only on (or be more prevalent on) the windward side of the tree and the windward side of erect shoots**

- probably early stages of **winter injury** (see part III, section 4.6.2) and likely to result in die-back.

31a **Affected leaves reduced to transparent papery brown skin or skeleton**

- **insects**. There are common leaf "skeletonizers" of willow (see part III, section 6.11) and poplar; other broadleaves occasionally suffer similar damage.

31b **On birch; on close inspection, leaves have numerous, coalescent, yellow to dark brown spots with a yellowish orange or brown powdery coating on the underside**

- possibly **rust**; see part III, section 6.3. Rusts are normally unmistakeable, but the disease can be so severe that leaves look scorched or frosted.

Key Section continues

31c On poplar; crown defoliated from base; in severe cases only a tuft of foliage remains at the top of the crown by late summer/early autumn

- possibly ***Marssonina*** **leaf-spot**; see part III, section 6.7.

31d On elm; foliage in a patch in the crown (or over an extensive area) yellowing and browning; under the live bark of affected shoots outer wood has longitudinal brown or purple streaks

- **Dutch elm disease**; see part III, section 6.9.

31e None of these genera, or not with these symptoms

- complete discoloration of leaves can be caused by several diseases and a variety of abiotic injuries which cannot be separated on field characters alone.

32a On a conifer **47**

32b On a broadleaf; pustules yellow or orange at first but may be replaced by brown sporing pustules later in the season, when the leaves may senesce prematurely;

- likely to be a **rust**; the most common ones are noted in part III, section 6.3.

32c On oak, sycamore or hawthorn; white pustules or white powdery deposit

- likely to be **mildew**.

33a On a conifer **34**

33b On a broadleaf **40**

34a On Scots pine; symptoms in autumn to spring

- probably ***Lophodermium***, but ***Lophodermella*** a possibility; see part III, sections 4.3 and 4.4.

34b On Scots or Corsican pine; symptoms in late summer

- probably ***Lophodermella***, but ***Lophodermium*** a possibility; see part III, sections 4.4 and 4.3.

34c On Norway spruce, Sitka spruce or lodgepole pine **35**

34d On larch

- probably ***Meria laricis***; see part III, section 5.1.

34e On Douglas fir **39**

34f On young noble fir; current needles develop reddish brown bands or become completely brown, and possibly withered, during summer; may affect few or many needles variously distributed on shoots

- **summer needle browning, cause unknown**, a not infrequent problem in Christmas tree crops.

35a **Tiny yellow flecks (less than 1 mm) all over needle, sometimes coalescing to give a yellow tip**

- common in Sitka spruce, occasional on lodgepole pine, on high elevation, exposed, or coastal sites where its cause is unknown; elsewhere, on spruce, this might indicate the first stages of attack by mites (**conifer spinning mite** or ***Nalepella*** spp.) and expert advice should be sought if Christmas tree crops are at risk.

35b **Not so; on spruce** **36**

36a **Needles and shoots with sticky deposits and small (0.5–1 mm) green aphids present**

- ***Elatobium;*** see part III, section 3.2.

36b **Without these features** **37**

37a **From September to following spring; youngest needles <u>only</u> with a central pinky-brown to red-brown band without sharp edges, the base and tip of the affected needles remaining green. On individual shoots, all needles may be affected, or those at the base and tip of the shoot may remain undamaged**

- **autumn frost**, see part III, section 3.1. In affected stands, some trees are likely to have all their youngest needles discoloured red-brown.

37b **Affected needles with large (more than 1 mm) yellow to brown, randomly positioned and sharply delineated blotches or bands** **38**

38a **In summer to winter; broad yellow or brown bands on current foliage only; older needles unaffected; unlikely to affect more than a few trees at any one location**

- ***Chrysomyxa* rust**; pustules eventually develop in late summer–autumn (***Chrysomyxa rhododendri***) or in late spring–early summer (***C. abietis***); see part III, section 3.6.

38b **In spring to early summer; yellow to brown spots, blotches and bands on the previous year's foliage; affected needles falling; damage may spread just into the basal 2–3 cm of the current shoot; likely to affect many trees**

- ***Elatobium***, see part III, section 3.2.

39a **Brown spots, blotches or pustules**

- ***Rhabdocline pseudotsugae.***

39b **Needles with blotchy yellow discoloration; with white woolly tufts at first and noticeably kinked**

- **woolly aphid (*Adelges (Gilletteella) cooleyi*).**

40a On oak

- possibly ***Apiognomonia errabunda***; see part III, section 6.4.

40b On sycamore **41**

40c On poplar or willow **42**

40d On birch; small yellow spots with a yellowish orange powdery coating on the lower surface

- probably **rust**; see part III, section 6.3.

41a Circular black spots with yellow halo, becoming thick and tough in late summer

- **tar spot**; see part III, section 6.5.

41b Not so

- there are a number of possible causes of leaf-spot and browning that may require laboratory examination to diagnose; see part III, section 6.5.

42a Small yellow spots with a yellowish orange powdery coating on the upper or lower surface

- probably **rust**; see part III, section 6.3.

42b Dark spots or blotches

- likely to be **anthracnose (*Marssonina* sp.)** or **scab (*Venturia* sp.)**. These require microscopic examination to identify; see part III, sections 6.7 (poplar) or 6.11 (willow).

43a On beech; affected parts of the leaves papery, transparent and hollow when held up to the light

- ***Rhynchaenus* leaf miner**; see part III, section 6.2.

43b On oak or beech; leaf spots or blotches, sometimes large areas of discoloration, at margin or forming a pattern between the veins

- possibly ***Apiognomonia errabunda***; see part III, section 6.4.

43c On lodgepole pine on peaty ground; tips of needles yellow grading into green without a sharp boundary, progressing to complete yellowing on older foliage

- probably **potassium deficiency**; see part III, section 4.1.

43d None of these

- the symptom is consistent with certain types of **nutrient deficiency, climatic damage (wind desiccation, frost, drought) or chemical damage.** It is not practicable to attempt to separate these on symptoms alone and, if the damage is significant, specialist advice is probably warranted. **Marginal browning of sycamore leaves** is sufficiently common on some trees in summer to deserve mention; the cause is unknown but the symptoms probably reflect some form of desiccation injury (see part III, section 6.5).

44a On *Prunus* or rowan; leaves rolled but not thickened, with sticky deposits and signs of insect activity inside rolled-up leaves

- **aphids**; there are several species that might be involved.

44b On Douglas fir; tuft of white wool associated with a yellow spot and a kink in the needle

- woolly aphid (*Adelges (Gilletteella) cooleyi*).

44c Not so **45**

45a Leaves or needles with protuberances or outgrowths from an otherwise normal structure **46**

45b Leaves cupped, curled, puckered or with blisters which deform both upper and lower surface of the leaf blade ... **49**

46a On conifer needles **47**

46b On broadleaves **48**

47a On Scots or Corsican pine; white or orange outgrowths up to 5 mm long from yellow spots

- *Coleosporium* **rust**; see part III, section 4.2.

47b On Sitka or Norway spruce; white or orange outgrowths, 2–3 mm long, from yellow or brown spots

- *Chrysomyxa* **rust**; see part III, section 3.6.

47c On Douglas fir; brown pustules

- *Rhabdocline pseudotsugae*.

48a On rowan; horn-like outgrowths from leaflets

- *Gymnosporangium cornutum*.

48b On sycamore; raised black "tar spots"

- **tar spot**; see part III, section 6.5.

48c On oak

- **insect galls.** There is such a variety of these associated with oak that they are not further keyed here.

48d On other broadleaves; pimples, warts, or green/white felty cushions on the leaf blade; may be very dense and may turn reddish

- probably **eriophyid mites**; see part III, section 6.2.

49a **On alder; leaves thickened and puckered or blistered; young green shoots may also be thickened and wavy**

- *Taphrina tosquinettii.*

49b **On poplar; smooth blisters, yellow on concave surface; convex surface may remain green; whole leaf blade may be deformed**

- *Taphrina populina.*

49c Not so **50**

50a **Leaves with small yellowish, reddish, or brownish, felty blisters**

- possibly **eriophyid mites**; see part III, section 6.2

50b **Not so**

- there are several fungi, insects and mites that can cause foliage deformation on forest broadleaves but they are not further keyed here as they are uncommon, or difficult to identify, or both. However, some translocated herbicides can induce deformed growth (part III, section 1.3.1) and this possibility should be borne in mind if these symptoms occur on young trees.

51a **On broadleaves; leaves very reduced (only 1 or 2 cm long), with a feathery appearance if normally serrate or pinnate; affected leaves may grow in tufts and have yellow or white discoloration of the margin or whole lamina, or may develop a reddish colour**

- these symptoms can be induced by certain **herbicides** in the year following treatment (and sometimes for several years thereafter) and, exceptionally, by rare types of **climatic damage**.

51b **On broadleaves; leaves not so severely reduced; may be yellowish but, if so, the discoloration is not confined to the margin**

- could indicate **drought** or a **root problem**. In large trees, the latter possibility would merit further investigation because of the safety implications.

51c **On Sitka spruce; needles unusually short and closely spaced but not discoloured; plants in at least the third year after planting**

- probably **phosphorus deficiency** (see part III, section 3.5) but **herbicide** damage cannot be ruled out if a translocated herbicide has been used nearby. **Note** that it is common for shoots to be short, with short closely spaced needles, in the year of planting (and sometimes in the year thereafter) as a reaction to the shock of transplantation.

51d **On Sitka spruce; needles very short, and may become gradually shorter towards the shoot apex; with bright yellow or white tips, or completely bright yellow**

- **translocated herbicide.**

52a Damage affects buds only, or buds plus shoot tips, of spruce **53**

52b Damage consists of shoot proliferation, abnormal growth, abnormal branching, or "witches' brooms" with or without die-back **54**

52c Damage consists only of resin bleeding from the main stem of conifers **106**

52d Not these; damage involves branch die-back, or lesions or cankers on branches **55**

53a On Norway spruce; high in the crown of pole-stage or older trees; death of buds and shoot apices; affected buds swollen and distorted with a black encrustation; may be associated with production of short, hooked shoots and with a zig-zag branching pattern

- probably ***Cucurbitaria* bud blight**; see part III, section 3.7.

53b On Sitka or Norway spruce; death of buds and shoot apices without swelling; on young trees or lower branches of older trees

- possibly **frost**; see part III, section 3.1

54a Branches fused and grotesquely flattened; may be curled into shapes like the antlers of fallow deer; affecting single trees (rarely two or three in a stand)

- **fasciation**; see part III; section 1.6.

54b On Norway spruce; high in the crown of pole-stage or older trees; short, hooked shoots or shoots with zig-zag branching; may lead to clumps of deformed, crabbed shoots; associated with swelling and death of buds, and a black encrustation on affected buds;

- probably ***Cucurbitaria* bud blight**; see part III, section 3.7.

54c On alder; young shoots and leaves thickened and deformed; leaves puckered or blistered

- *Taphrina tosquinettii.*

54d On birch, *Prunus*, or hornbeam; abnormal proliferation of branches to give a dense bunch of twigs (a "witches' broom"); live brooms may have abnormally sized or coloured leaves

- *Taphrina* sp. This genus of fungi has several members that can induce deformation in shoots and leaves on broadleaves.

Key Section continues

54e Not fitting any of these descriptions

- there are several insects, mites and fungi that can cause abnormal growth in forest trees but they are not further keyed here as they are uncommon, or difficult to identify, or both. Some translocated herbicides can induce abnormal branching or deformed growth (part III, section 1.3.1) and this possibility should also be borne in mind.

55a Severe crown die-back progressing to death of trees or seeming likely to **85**

55b Die-back present but not severe (affecting scattered shoots or less than 25% of crown in a patch), or branches have lesions or cankers without die-back **56**

56a Crop recently fertilized or treated with herbicide; symptoms developing more or less simultaneously on several trees; foliage browning and die-back of young (1–3 year) shoots

- **chemical damage** is a strong possibility and might be difficult to eliminate; the distribution of damage in the crop could provide further evidence but specialist advice might be required. See part III, sections 1.2 and 1.3.

56b No chemical treatment in affected crop **57**

57a Single tree or small group; death of top or side branches associated with long vertical scar or strip of dead bark on main stem

- possibly **lightning**; see part III, section 1.4.5.

57b No scarring **58**

58a On a conifer **59**

58b On a broadleaf **76**

59a On spruce **60**

59b On pine **64**

59c On larch **72**

59d On Douglas fir **75**

59e On grand fir; browning and die-back in late winter or spring on shoots of the previous year; may involve death of buds and non-lethal shoot injury expressed as reduced extension of current shoots; may affect scattered shoots or the whole windward side of the crown in severe cases

- **winter injury**; see part III, section 1.4.2.

Key Section continues

59f **On *Tsuga* or *Thuja*; general browning and some die-back in crop**

- these symptoms are consistent with **drought** (part III, section 1.4.4) or **winter injury** (part III, section 1.4.2), to which these genera are susceptible. In *Thuja*, interior branches and foliage may be particularly affected by drought. If neither explanation fits the circumstances, specialist advice may be required.

60a **On Norway spruce over 10 years old; browning and die-back affecting the whole crown, or the windward side, or patches in the mid-crown. On affected shoots, browning of foliage starts at shoot tips (fig 31)**

- probably the early stages of **top-dying** or **drought** damage; see part III, sections 3.4 and 1.4.4.

60b **Not so; on Sitka or Norway spruce (but note that from here onwards the key sections refer principally to Sitka spruce) 61**

61a **In spring to summer; many soft current shoots killed, "tasselled", or deformed; tips of shoots of previous year, and buds, may also be killed**

- **frost** is the most likely cause but there are other possibilities; see part III, section 3.1.

61b **No tasselling; death of shoot tips and buds (including leading bud) of young trees**

- probably **autumn frost** but early (preflushing) **spring frost** is also a possible explanation; see part III, section 3.1.

61c **Fitting neither description 62**

62a **Affected trees on peaty ground; severe yellowing or loss of older needles, with current needles remaining green and needle length normal; over the whole crown or in patches; die-back affecting subsidiary shoots on side branches while leading shoots usually survive**

- probably **potassium deficiency**; see part III, section 3.5.

62b **Foliage present and of normal colour on live parts of shoots 63**

63a **Death of scattered shoots in mid or upper crown of pole-stage and older trees**

- probably **winter injury** or **wind abrasion**; see part III, sections 1.4.2 and 1.4.3.

Key Section continues

63b **Affected trees with abnormally short and closely spaced (but not discoloured) needles and extremely poor extension growth on live shoots; trees in at least their third year after planting**

- probably **phosphorus deficiency** (part III, section 3.5) but **herbicide damage** cannot be ruled out if a translocated herbicide has been used nearby. **Note** that it is common for shoots to be short, with short closely spaced needles, in the year of planting (and sometimes in the year thereafter) as a reaction to the shock of transplantation.

64a **Die-back of Scots pine associated with wholesale foliage browning in late winter or spring; may affect whole plantations or groups of Scots pine in a matrix of other species; symptoms remain confined to older needles as live shoots flush**

- likely to be severe ***Lophodermium*** infection but with the possible involvement of other agents (see part III, sections 4.3 and 4.6.5); **winter injury** (part III, section 4.6.2) is a less likely explanation.

64b **Not so** **65**

65a **Dead or dying shoots up to 1 year old having a hollow centre and/or a resinous bore-hole near the apex**

- **pine-shoot beetle (*Tomicus piniperda*)**; see part III, section 4.6.1.

65b **Shrunken or swollen lesions or cankers present** **66**

65c **Die-back of 1–3 year shoots without these features** **68**

66a **On lodgepole pine; cankers often resinous and blackened**

- probably ***Crumenulopsis* canker**; see part III, section 4.6.6.

66b **On Scots pine; very open cankers with wide, flared margins; usually on side-branches; despite the deformation, not often associated with die-back, nor abnormally resinous**

- **cause unknown** but sometimes associated with the fungus *Sarea difformis*.

66c **On Scots or Corsican pine; resinous cankers, swollen or sunken but not with widely flaring margins** **67**

67a **Whitish yellow to pale orange pustules present (in early summer) on swollen bark around the lesion**

- ***Peridermium***; see part III, section 4.5.

67b **Lesion with resinous, swollen, roughened bark; on larger branches bark may be cracked at the centre of the canker**

- probably ***Peridermium*** but this can be difficult to distinguish from ***Crumenulopsis* canker**; see part III, sections 4.5 and 4.6.6.

Key Section continues

67c **Resinous sunken lesion or canker with callus growth but a sunken centre, possibly with exposed wood**

- probably *Crumenulopsis* **canker** but *Peridermium* is a possible explanation; see part III, sections 4.6.6 and 4.5.

68a **Discoloration of needles clearly extends further down the shoot (i.e. from the bud) than death of bark**

- **winter injury** (part III, section 4.6.2) or, on Corsican pine, **frost injury** (part III, section 4.6.7).

68b **This feature not apparent; death of bark precedes or coincides with discoloration of needles, or the relative positions are not clear 69**

69a **On lodgepole pine 70**

69b **On Corsican pine 71**

69c **On Scots pine; bud-death and die-back developing in late winter and spring**

- *Brunchorstia*, *Crumenulopsis*, or (less likely) **winter injury**. The first two are difficult or impossible to tell apart but *Brunchorstia* is the most likely if die-back affects more than a few shoots in the lower crown. See part III, sections 4.6.2 and 4.6.4–4.6.6.

70a **Bud or very tip of shoot dead with yellow and brown needles attached**

- probably *Ramichloridium*; see part III, section 4.6.3.

70b **Die-back and browning more extensive on affected shoots**

- **winter injury** or *Ramichloridium*; see part III, sections 4.6.2 and 4.6.3.

71a **On trees up to 1 m; death of extending buds or shoots in spring/early summer; affecting many plants; softer shoots may be "tasselled"; shoot die-back accompanied by red–brown discoloration of foliage**

- likely to be **spring frost injury**; see part III, section 4.6.7.

71b **Without this combination of symptoms or on larger trees; bud-death and/or die-back developing in autumn to spring**

- *Brunchorstia*, *Crumenulopsis*, or (less likely) **frost injury**. These are difficult or impossible to tell apart on field symptoms, but *Brunchorstia* is by far the most likely if die-back is extensive; see part III, sections 4.6.4–4.6.7.

72a On European or hybrid larch; affected branches have resinous lesions or cankers with fruit bodies; latter 2 or 3 mm across, scattered or in a circle on the canker face; either disc-shaped with an orange yellow centre and white rim, or (in dry weather) closed into a white ball

- **larch canker**; see part III, section 5.2. **Note** that, in their early stages, cankers are barely noticeable except perhaps as flattened areas with slight resin bleeding.

72b No cankers with fruit bodies **73**

73a Die-back of isolated, scattered 1–3 year old shoots; particularly noticeable in spring due to the browning and shrivelling of recently flushed dwarf shoots; death usually occurs before the extension shoots flush **74**

73b Die-back prevalent on lower 3 m or so of trees; death of tips of previous year's shoots in spring; may be accompanied by "tasselling" of new growth

- probably **frost injury**; see part III, section 1.4.1.

73c On European or hybrid larch; die-back associated with white woolly deposits on foliage of whole tree; needles blotched, blackened and sticky

- **"larch die-back"** caused by the **larch woolly aphid**; see part III, section 5.3.

74a Bore-hole present at junction of dead and live tissue

- **larch shoot moth** (*Argyresthia laevigatella*); see part III, section 5.5.

74b No bore-hole; shrunken lesion or blob of resin at the junction of dead and live tissue

- probably ***Phacidium coniferarum*** or ***Cytospora* sp.** (part III, section 5.5) but, on European or hybrid larch, possibly **larch canker** (part III, section 5.2).

75a On trees up to 2 m; death of buds or die-back developing in late winter, spring or early summer

- **frost** or **winter injury** which may have been extended by ***Phomopsis* disease**; see part III, section 1.4.1.

75b On larger trees; browning and die-back in late winter or spring on shoots of the previous year; may involve death of buds and non-lethal shoot injury expressed as reduced extension of current shoots; may affect scattered shoots or the whole windward side of the crown in severe cases

- **winter injury**; see part III, section 1.4.2.

76a On beech **77**

76b On oak **78**

76c On elm **79**

76d On poplar or aspen **81**

76e On willow **82**

76f On *Prunus* **83**

76g On ash **84**

76h On birch, decline of trees associated with fungal brackets on the main stem

- see part III, section 6.12.

77a Sunken lesions or cankers present; leading to die-back of small (less than 25 mm diam.), isolated branches (occasionally larger ones)

- probably ***Nectria* canker**.

77b Not so; trees over 10 years old **99**

78a Severe foliage damage present; affected leaves covered in white deposit, discoloured dingy yellow or brown and may be slightly puckered

- **oak mildew**. This disease normally only affects foliage but severe attacks can lead to die-back.

78b Severe foliage damage present; affected leaves with large brown blotches, or completely brown, but without a white deposit

- probably ***Apiognomonia errabunda***; see part III, section 6.4. Confirmation would require laboratory examination.

79a Die-back of isolated twigs and small branches associated with cankers (cankers may be present on larger branches without causing die-back)

- probably ***Plectophomella* or *Nectria* canker**. If pin-head sized fruit bodies are present on the face of the canker, go to **80**.

79b Patches of the crown affected by yellowing, wilting, defoliation and die-back; on shoots with wilting foliage and still-live bark, the outer wood has brown or purple streaks

- **Dutch elm disease**; see part III, section 6.9.

80a **Pin-head sized, dark brown or black, fruit bodies present on the canker**

- *Plectophomella concentrica*.

80b **Pin-head sized, pink or red, fruit bodies present on the canker**

- *Nectria* sp.

81a **Affected shoots, branches or stems having erupting or sunken cankers which, in wet weather in spring, may exude whitish slime through splits in the bark**

- probably **bacterial canker**; see part III, section 6.7.

81b **Die-back of young shoots associated with black lesions; lesions originate at the points of attachment of blackened, shrivelled leaves in the early stages but affected shoots quickly become completely blackened, and hooked or shrivelled; symptoms may be severe on aspen**

- ***Venturia***; occasional on aspen, uncommon on other poplars; see part III, section 6.7.

82a **Elliptical brown or black lesions on young shoots, not all of which develop die-back; brown leaf spots also present; die-back, if present, affecting a few shoots only**

- probably ***Marssonina salicicola***, especially if a weeping variety is affected, but microscopic examination is necessary to distinguish this from **scab** and **black canker** diseases (next key section) with confidence; see part III, section 6.11.

82b **Affected leaves and young shoots blackened and shrivelled; black lesions or cankers present on shoots; die-back may become severe even on large trees (especially of crack willow)**

- probably **scab** or **black canker** disease but microscopic examination is necessary to identify these and to distinguish them from ***Marssonina salicicola*** with confidence; see part III, section 6.11.

82c **Die-back preceded by red-brown discoloration and wilting of leaves on the affected branches; wood of affected branches and the main stem below them with a watery brown stain that turns reddish on exposure**

- **watermark disease**; common only on cricket bat willow but some commercial Dutch clones are also susceptible; see part III, section 6.11.

83a **Affected branches or main stem with sunken, gummy lesions or cankers; roots and stem base remain alive on trees with die-back**

- probably **bacterial canker**; see part III, section 6.6.2.

Key Section continues

83b **Die-back of portions of the crown preceded by leaden or silvery discoloration of foliage; wood of dying branches stained brown; small bracket-like fruit bodies with a purple underside on dead bark**

- **silverleaf disease** (***Stereum purpureum***); see part III, section 6.6.2.

83c **Die-back of young shoots or spur shoots associated with withering of leaves or blossom**

- either **bacterial canker** or ***Sclerotinia* disease** which are difficult to separate on field characters. See part III, section 6.6.2.

84a **Swollen "knots" or erumpent, hard, lumpy cankers present**

- probably **bacterial canker**. The currently correct Latin name of the pathogen is *Pseudomonas syringae* subspecies *savastanoi* pathovar *fraxini*, but some books may omit one or more from the list.

84b **Sunken or "target" cankers present (i.e. cankers with regular concentric ridges of callus around them)**

- probably ***Nectria* canker** but **bacterial canker** is still a possible explanation.

85a **Death of single trees or small groups consistently located on roadsides** **86**

85b **Trees not consistently on roadsides** **87**

86a **Base(s) buried by road operations**

- **asphyxiation**; may be compounded by secondary attacks by opportunistic pests and pathogens.

86b **Adjacent to, or just downslope of, a road grit dump**

- **salt;** see part III, section 1.3.2.

86c **Not so**

- **dumping of toxic material** from the road is a possible explanation; the death of ground vegetation would provide further evidence. Otherwise, assume the association with the road is purely coincidental and follow the rest of the key from 87.

87a **Single tree or small group affected; death, death of top, or death of major limbs associated with long vertical scar or strip of dead bark on main stem; if a group of trees is affected, some fringing live trees may have dead limbs facing into the core of dead trees**

- possibly **lightning**; see part III, section 1.4.5.

87b **No scarring** **88**

88a **Dead/dying trees in group(s), swathe(s) or patch(es) consistently associated with watercourses or drainage pattern; live, symptomatic trees with dead/dying roots but live main stem**

- most likely to be **waterlogging**. If this can be ruled out, expert advice would be required to investigate other possibilities such as **groundwater pollution** or ***Phytophthora* disease**. See part III sections 1.5.1, 1.5.2, 2.1 and 6.13.

88b **Dead/dying trees in a group or patch; signs of scorching or charring on stems; burnt wood or charcoal throughout site; fringing live trees may have basal scars**

- **fire.**

Note **on a and b.** The signs and effects of recent waterlogging or fire are unmistakeable but, when vegetation has regrown, these are causes of "group killing" that can be easily overlooked.

88c **No association with drainage or fire** **89**

89a **Conifers affected** **90**

89b **Broadleaves affected** **97**

90a **Dead/dying trees with fruit bodies ("brackets") at the groundline; affected trees scattered or in small groups; brackets (figs 21 and 22) with a dark brown upper surface, creamy white margin and undersurface, the latter with numerous tiny pores; on small trees these may be reduced to small, creamy white and pale brown lumps (fig 23)**

- ***Heterobasidion annosum***; see part III, sections 2.1 and 2.3.

90b **Dead/dying trees with fans or sheets of cream or white fungal tissue (about the thickness of paper but softer) in the bark or between bark and wood at the base of the stem; affected trees scattered or in small groups; usually associated with resin bleeding from lower stem**

- ***Armillaria***; see part III, sections 2.1 and 2.2. **Note** that this fungus will invade trees killed by other agents. For a sound diagnosis, mycelium should be present in all symptomatic but still living trees; an examination restricted to dead trees is not enough. Though fragile, *Armillaria* mycelium is quite substantial; very thin or wispy white fungal growth is unlikely to belong to *Armillaria*.

90c **Without brackets or mycelium** **91**

91a **Dead/dying trees in one or more discrete groups less than 100 m across; groups centred on the sites of bonfires; in summer (June to September) dark red or chestnut brown, irregularly rounded, hollow fruit bodies (fig 24) up to 10 cm across appear on the ground in the group or around its edges; there may be heavy resin bleeding from the lower main stem**

- *Rhizina*; see part III, sections 2.1 and 2.4.

Note In the absence of fruit bodies, the combination of group killing and fire sites is enough to diagnose *Rhizina* with reasonable certainty.

91b **Not so** **92**

92a **Norway spruce over 10 years old affected; symptoms a mixture of patchy crown browning proceeding to complete browning, die-back and death; browning of foliage starts at shoot tips (fig 31) and, even in advanced stages of crown decline, the roots remain alive**

- **top-dying** or, less likely, **drought**; see part III, section 3.4.

92b ***Thuja*** **or** ***Tsuga*** **affected; site recently subjected to drought; trees suffering general browning and leader die-back; damage may be general or patchy depending on topography**

- possibly direct injury by **drought**, to which these genera are rather susceptible; in *Thuja*, damage may be noticeably more severe in the interior of the crown. See part III, section 1.4.4.

92c **Douglas fir up to 1 m affected; die-back of leader and/or side shoots developing in late winter, spring or early summer; lower stem and roots remain alive; trees without die-back, and live shoots on trees with die-back, of normal appearance**

- possibly **Phomopsis disease** following **frost** or **winter injury**; see part III, section 1.4.1.

92d **Pine affected** **93**

92e **Larch affected** **96**

93a **Lodgepole pine over 10 years old affected; die-back of shoot tips throughout the crown; roots remain alive at an advanced stage of crown die-back; on the most recently affected shoots, damage consists of a bunch of yellow and brown needles at the shoot tip; die-back often starts, or is worse, in the lower crown, but can lead to decline and death**

- probably ***Ramichloridium*** **die-back**; see part III, section 4.6.3.

93b **Corsican pine affected** **94**

93c **Scots pine over 10 years old affected** **95**

94a On small (<1 m) trees; shoots killed in spring/early summer; affecting many trees; softer shoots may be "tasselled"; die-back accompanied by reddish brown discoloration of foliage

- likely to be **spring frost injury**; see part III, section 4.6.7.

94b On larger trees or damage not as described **95**

95a Few to occasional trees affected; all have a black, resinous canker on the main stem below the symptomatic part of the crown (these are often still visible on long-dead trees)

- probably ***Peridermium* canker**; see part III, section 4.5. Note that this disease occurs, but is rare, on Corsican pine

95b Many trees, or whole crop, with severe foliage browning and die-back proceeding to death; dying trees still with live roots (though they are likely to deteriorate rapidly once the trees are dead); in the least affected trees symptoms may be confined to the lower crown

- likely to be ***Brunchorstia* disease** though, in the case of Scots pine, possibly compounded by ***Lophodermium*** infection and ***Tomicus*** attack (part III, sections 4.6.4 and 4.6.5). However, on soils with a high chalk or limestone content, severe **lime-induced chlorosis** might instigate rather similar symptoms.

96a Die-back of upper crown or entire crown; affected trees have insect galleries beneath the bark; in large trees only the top of the crown may be affected and then the galleries may be confined to the upper main stem; resin bleeding may occur from insect entry holes

- ***Ips cembrae***; see part III, section 5.4.

96b On European or hybrid larch; foliage with white woolly deposits; needles blotched, blackened and sticky; die-back of small and large branches occurring all over the crown

- "**larch die-back**" caused by the **larch woolly aphid**; see part III, section 5.3. **Note** that serious infestations are often associated with **larch canker** (next key section) and it may be impossible to decide which is the more damaging in a particular situation.

96c On European or hybrid larch; dead and dying branches with resinous lesions or cankers; lesions not associated with insect tunnels as in 96a

- **larch canker**; see part III, section 5.2.

96d Without these symptoms; site recently subjected to drought, trees suffering general browning and leader die-back; damage may be general or patchy depending on topography

- possibly direct injury by **drought**, to which larches are rather susceptible; see part III, section 1.4.4.

97a On elm; parts of the crown, or the whole crown, affected by yellowing, wilting, defoliation and die-back; on shoots with wilting foliage and still-live bark, the outer wood has brown or purple streaks; in advanced stages of crown die-back, the roots and stem base remain alive

- **Dutch elm disease**; see part III, section 6.9.

97b On birch; decline of trees associated with fungal brackets on the main stem

- see part III, section 6.12.

97c On *Castanea*; decline or die-back of crown associated with death of roots and inverted-V-shaped lesions at the stem base; wood of dead/dying parts may be stained blue-black and inky fluid may exude from affected parts

- likely to be **ink disease**; see part III, section 6.13.

97d Not fitting any of these descriptions **98**

98a On beech over 10 years old **99**

98b On poplar or aspen **100**

98c On willow **101**

98d On *Prunus* **102**

98e On sycamore **103**

98f Not these **104**

99a Main stem covered with a white deposit and having "tarry spots" (patches of bark weeping black fluid) or patches of dead bark; crown symptoms may proceed through thinning, yellowing, die-back and decline to death, but roots remain alive until the crown dies; affected stands often contain broken trees

- **beech bark disease**; see part III, section 6.8.

99b No white deposit; weeping patches or areas of dead bark present on main stem

- possibly **beech bark disease** from an old coccus attack, or a stress-related disorder (**drought** being the most likely). See part III, section 6.1. **Note** that drought-related declines may continue for several years after the drought that initiated them.

Key Section continues

99c **No stem injury apparent; diffuse crown thinning and die-back, especially at top of crown; may be associated with production in autumn of large fan-shaped, pliable, poroid fruit bodies at the stem base or with large, woody perennial brackets on the stem base**

- possibly a **root disease** (part III, section 6.1); the fruit bodies are likely to be *Meripilus giganteus* or *Ganoderma* sp. (brackets) but other less conspicuous fungi can induce a similar condition. Old beech are especially prone to root disease and can become dangerous as a result; crown deterioration need not be severe for trees to be unstable.

100a **Die-back associated with erupting or sunken cankers which, in wet weather in spring, may exude whitish slime through splits in the bark**

- probably **bacterial canker**; see part III, section 6.7.

100b **Die-back of young shoots associated with black lesions; lesions originate at the points of attachment of blackened, shrivelled leaves in the early stages but affected shoots may quickly become completely blackened, and hooked or shrivelled; symptoms may be severe on aspen**

- ***Venturia***; occasional on aspen, uncommon on other poplars; see part III, section 6.7.

100c **Not with these symptoms** **104**

101a **Leaves and young shoots blackened and shrivelled; black lesions or cankers present on shoots; die-back may be severe even on large trees (especially of crack willow)**

- probably **scab** or **black canker** disease but microscopic examination is necessary to identify these and to distinguish them from ***Marssonina salicicola*** with confidence; see part III, section 6.11.

101b **Die-back preceded by red-brown discoloration and wilting of leaves on the affected branches; wood of affected branches and the main stem below them with a watery brown stain that turns reddish on exposure**

- **watermark disease**; common only on cricket bat willow but some commercial Dutch clones are also susceptible; see part III, section 6.11.

101c **Not with these symptoms** **104**

102a **Die-back associated with sunken gummy lesions on branches or main stem; roots and stem base remain alive**

- probably **bacterial canker**; see part III, section 6.6.2.

Key Section continues

102b Die-back of portions of the crown preceded by leaden or silvery discoloration of foliage; associated with brown staining in wood of dying branches; small bracket-like fruit bodies with a purple underside on dead bark; roots remain alive

- **silverleaf disease** (***Stereum purpureum***); see part III, section 6.6.2.

102c Not with these symptoms **104**

103a Die-back of crown associated with blistering and flaking of the bark; a dry, sooty, black, powdery deposit present under detaching bark

- possibly **sooty bark disease** (see part III, section 6.10) but this disease is easily confused with other problems; it is unlikely to be found in northern England or Scotland.

103b No dry sooty deposits but die-back associated with cankers or flaking bark on the main stem or limbs

- possibly **drought**; see part III, section 1.4.4. **Note** that the effects of drought injury in sycamore may not become apparent for some years.

103c Not with these symptoms **104**

---

104 GENERAL NOTE. Die-back in large, old forest broadleaves is difficult to diagnose. If key sections 104 and 105 are read in conjunction with the general notes in part III, section 6.1, they may help in understanding the nature of the problem, though they will not point to a precise cause.

104a Trees dying back, main stem alive but roots dead or dying **105**

104b Not so

- likely to be an unusual (or unusually developed) shoot disease or a form of abiotic damage.

105a Fans or sheets of cream or white fungal tissue in the bark or between bark and wood at the base of dying trees; the fungal tissue (mycelium) should be about the thickness of paper but softer

- ***Armillaria*** root disease is involved but, in broadleaves, may not be the primary cause.

105b No substantial white mycelium as described in 105a

- **a root problem** other than *Armillaria* is indicated.

106a **Inner bark mined or excavated; resin tubes may be present**

- **bark beetle** infestation. Such attacks are generally stress-related but could involve dangerous non-indigenous species; advice should be sought from the Forestry Commission where this is suspected.

106b **No physical damage to bark 107**

107a **Affected trees in distinct groups associated with the sites of fires; in summer (June to September) dark red or chestnut brown, irregularly rounded, hollow fruit bodies (fig 24) up to 10 cm across appear on the ground in the group or around its edges; resin bleeding from the lower main stem may be heavy**

- ***Rhizina*** (early stages of attack); see part III, sections 2.1 and 2.4.

107b **Not in groups or, if in groups, not associated with sites of fire 108**

108a **Site is second (or more) rotation; resinosis usually close to the ground line**

- likely to indicate root- and butt-rot by ***Armillaria*** or ***Heterobasidion annosum***. If the latter, butt-rot could be severe. See part III, sections 2.1, 2.2, 2.3 and 8.3.

108b **First rotation site; thinned**

- possibly rapid and severe development of root- and butt-rot by ***H. annosum***; see part III, sections 2.1 and 2.3.

108c **First rotation site; unthinned**

- still likely to indicate a **root problem** but an obscure one, or a common one affected by some unusual feature of site or crop management. In either case, specialist investigation may be needed.

PART III

NOTES ON COMMON DISEASES AND DISORDERS OF PLANTATIONS AND WOODLANDS

1. ABIOTIC DISORDERS AFFECTING SEVERAL SPECIES

The diagnosis of damage caused by non-living agents relies heavily on circumstantial evidence, a good knowledge of tree anatomy, and on the interpretation of timing and pattern. In plantations, most instances of abiotic damage fall into one of the following categories: planting failure, site-related problems, climatic injury or chemical injury.

1.1. Planting Failure

Planting failure is a commonly used expression for the failure of plants to become properly established after outplanting. The reasons for its occurrence may be poor plant quality, bad handling and storage procedures, inadequate site preparation, bad planting practice, poor weed control, or a combination of such factors. The problem usually shows up in the weeks or months following planting as foliage browning and die-back, generally associated with root mortality and poor development of live roots. Failure may affect only certain batches or species in mixed plantings and such patterns may help to pin down the origin of a handling problem.

A large proportion of affected plants may die but usually some recover, sometimes by producing shoots from the base. In their first season, survivors frequently show poor shoot extension with, in conifers, short, closely spaced needles. Die-back and death in the second or third year are unlikely to be solely attributable to planting failure, though plants with poor root systems are at risk from heavy weed competition and drought, and might continue to struggle for some years. Widespread browning and die-back in young plants beyond their first season, if not associated with girdling by insects or mammals, are most likely to be explained by climatic or herbicide injury.

When herbicide has been used in the year of planting, it may be impossible to distinguish the effects of herbicide misuse from those of planting failure. The former would certainly be suggested if the damage to crop plants were to coincide precisely with areas of most effective weed control. Further evidence might come from the occurrence of some symptoms specific to the chemical used, or from chemical analysis. However, even with positive evidence of herbicide injury, it might still be impossible to rule out planting failure as a contributory factor to the overall damage.

Planting failure may be exacerbated by drought or waterlogging, both of which may lead to mortality in groups or in patterns associated with topography. It is worth commenting that, in the uplands, drought is not often the primary cause.

1.2. General Remarks on Damage by Climate and Airborne Chemicals

Toxic chemicals often have similar immediate effects to those of damaging extremes of climate: green tissues turn brown or black and soft shoots shrivel or collapse. If damage to foliage is partial, discoloration is usually confined to needle tips or, in broadleaves, occurs in a regular pattern around the margin or between the veins. Severe injury by either climate or a toxin is likely to (though not bound to) occur over a relatively short space of time after which those trees that survive will produce new growth, or attempt to do so. Both forms of injury are likely to exhibit protection effects, such as less severe damage on the leeward side of trees (figs 2 and 6) or on lower foliage. Thus it may be impossible to distinguish chemical damage from climatic damage on the basis of symptoms alone. If chemical injury is suspected, further advice on the feasibility of laboratory analysis may be required.

Special care is necessary in diagnosing climatic or chemical damage in pure Norway spruce or pure Scots pine because of the prevalence of the disorder "top-dying" in the former and stress-related browning by *Lophodermium* needle-cast in the latter (see sections 3.4 and 4.3, respectively). Top-dying is so prevalent that if foliage browning and dead trees are evident in a pure Norway spruce crop in any situation, it would be extremely difficult to rule it out of consideration. In the case of needle-cast in Scots pine, symptom development is apparently enhanced by stress. Consequently, browning often develops on the windward side of trees and may look very like the effects of exposure or chemical "scorching".

Fig 2. Directional damage; winter injury on Scots pine.

1.3. Injury by Chemicals

1.3.1. Herbicides and Fertilizers

The symptoms of herbicide injury vary according to the chemical used, dosage, plant species and time of year. Moreover, products are regularly changed or improved so that it would be impractical to attempt a prescriptive list of symptoms. However, some of those associated with modern herbicides (though not in themselves providing proof of herbicide injury) are:

yellowing or bleaching of needle tips and leaf margins;

reduced needle or leaf size – on spruce affected by glyphosate, needles may get progressively shorter towards the shoot apex (fig 3); on ash and oak, leaves may be only a centimetre or so long and have a feathery appearance;

failure of live terminal buds to flush;

short recovery shoots formed in bunches.

Fig 3. Herbicide (glyphosate) injury on Sitka spruce. Note the yellowing and shortening of needles.

Injury by a forest herbicide may be suggested by the pattern of weed control, for example if crop damage coincides with complete removal of weeds or with unusually effective or persistent control (fig 4). However, the symptoms caused by non-fatal doses of some translocated materials (such as glyphosate and imazapyr) may appear in the season following application when the level of weed control is no longer so obvious. In Christmas tree plantations, where rigorous weed control is normal, patterns are less likely to offer clues. Damage following the drift of herbicide from adjacent agricultural ground – a much less common cause of injury than is popularly supposed – usually has a directional pattern and is likely to affect other vegetation in addition to crop trees.

Fig 4. Herbicide (hexazinone) injury on Sitka spruce. The "scorched earth" level of weed control around the damaged tree in the foreground is a useful diagnostic feature.

Chemicals such as urea and potassium chloride, which provide the nitrogen and potassium in many composite fertilizers, may be extremely damaging to trees. These materials not only scorch foliage and young shoots by contact but may also cause lesions and die-back on sizeable stems if granules lodge in leaf axils or branch bases. Overdoses reaching the ground can cause injury by root uptake. In the dormant season, sub-lethal injury to shoots or buds sometimes leads to late flushing and the production of short, weak shoots in the following spring. However, the same effects can follow climatic damage.

1.3.2. Other Types of Chemical Injury

Damage in plantations by airborne chemicals, other than by misuse of fertilizers or herbicides, is rare and extremely unlikely to occur other than in the following situations:

a. near (within 1 or 2 km) an industrial plant or other potential source of airborne pollution (fig 5) – serious pollution incidents are often single episodes though chronic damage is possible;

b. within 5 km of the coast (fig 6) – trees right on the coast are likely to suffer chronic damage from sea salt spray, but further inland damage is only likely to occur as a single event associated with a major storm, particularly in summer;

c. alongside a major road heavily salted in winter – damage from road salt is normally restricted to 30 m or so from carriageways, but cases are known where wind-blown spray has carried some hundreds of metres across open ground from exposed motorways.

Fig 5. Atmospheric pollution injury in Norway spruce. The source was a burning coal-spoil tip approximately 500 m away.

Fig 6. Sea salt damage to Sitka spruce on rising ground within 1 km of the west coast of Kintyre.

Damage by road salt also occurs by root uptake of outwash from salt dumps (fig 7) or heavily salted carriageways. Sizeable trees can be killed adjacent to dumps, even small ones, whereas grass and other ground vegetation may show no effect. Damage by salt spray or outwash is not necessarily confined to winter; die-back and, in broadleaves, foliage symptoms may develop and extend in spring. Other kinds of soil-borne chemical injury are dealt with in section 1.5.2.

Fig 7. Group of dead Norway spruce adjacent to a road salt dump.

1.4. Climatic Injury

1.4.1. Frost

The commonest type of climatic damage in UK plantations is probably that caused by frost in autumn or spring to unhardened shoots and buds. In spring, the most characteristic forms of damage are shrivelling and browning of new shoots and leaves (fig 8); in conifers, the withering of small shoots or collapse of larger ones (fig 25) is commonly referred to as "tasselling". Although these symptoms can be induced by other agents that kill shoots or withhold water from them at a critical time, the scale, timing and synchrony of frost damage, both on individual trees and in plantations, usually leaves little room for doubting the cause. On young shoots that have not flushed, spring frost may kill buds, shoot tips and, in conifers, needles of the previous year, especially those at the shoot apex. This kind of damage may be indistinguishable in appearance from that caused by autumn frost or winter desiccation. On larger stems, frost can cause localized lesions ("frost cankers") or cracks but, as with winter cold injury, these forms of damage are rare in British woods and plantations other than in a few susceptible species – *Tsuga, Nothofagus, Eucalyptus* and non-native alders, for example.

Fig 8. Spring frost damage on beech. The undamaged trees are late-flushing species.

Shoots killed or injured by frost may be invaded by secondary pathogens which extend the original damage. This is best known in Douglas fir where the fungus *Phacidium (Potebniamyces) coniferarum* can attack young trees that have suffered frost injury (or winter injury, as described in the next section) and cause serious die-back.

This problem is commonly referred to as "Phomopsis disease" from the old Latin name of the fungus, *Phomopsis pseudotsugae*.

Susceptibility to frost damage is influenced by inherited characters and nutrition as well as by temperature, which in turn is influenced by topography. Thus, although frosting in plantations often causes uniform damage in hollows or over flats, this is not always the case as some individuals or small groups may be more resistant or susceptible than others.

Among plantation conifers, frost injury is of particular significance in Sitka spruce and Corsican pine and is therefore treated in more detail in sections 3 and 4.

1.4.2. Winter Injury

Injury as a direct result of low temperature in winter is a rare event in UK plantations but occurred on exotics in the winters of 1962/63 and 1981/82. *Nothofagus* suffered severe damage during the former and freezing injuries were recorded on a variety of conifers (including Corsican pine, lodgepole pine, *Pinus muricata, P. radiata,* Leyland cypress and Douglas fir) at several places in Britain after the extremely cold weather at the beginning of 1982. Although in many of the 1982 cases trees sustained fatal stem lesions, shoot and foliage symptoms were delayed until the following year (fig 9).

Fig 9. Young Corsican pine killed by low temperatures in January 1982. Photographed in May 1983; the far left tree was affected and may die but still has green needles and grew (albeit poorly) in 1982. Scots pine in mixture were unaffected.

Among the more common forms of winter injury exhibited by conifers are those that arise from desiccation when foliage and shoots lose water that cannot be replaced because roots are inactive in frozen or near-frozen soil. Particularly dramatic examples can occur in young or recently planted trees when spells of severe weather are broken by the sudden onset of strong, mild winds (fig 10).

Fig 10. Winter injury to two-year-old Sitka spruce showing desiccation of upper shoots. Photographed in the following June when the sheltered lower shoots, which had not been damaged, were flushing.

Fig 11. Winter injury to Douglas fir.

Fig 12. Winter injury to noble fir; buds on the shoot with discoloured foliage have either failed to flush or have produced weak new shoots.

Fig 13. Severe winter injury to Scots pine in mixture with Norway and Sitka spruce; the spruce suffered some foliage injury but the discoloured needles had been shed when the photograph was taken in the following August.

Winter injury to the crowns of thicket stage or older conifers is associated with symptoms ranging from needle browning to bud death and shoot die-back (fig 11). Foliage browning may also be followed by the delayed production of short, weak shoots in spring (fig 12). Sometimes only scattered shoots are affected and in such cases the damage may look very like the effects of disease. Particular care is necessary in lodgepole pine to distinguish winter injury from shoot die-back caused by *Ramichloridium pini* (sections 4.6.2 and 4.6.3). Winter injury is most common in lodgepole pine (and is

Fig 14. Winter injury to Scots pine; individual from the crop shown in fig 13.

further described in section 4.6.2) but it is also seen frequently in Douglas fir (fig 11) and *Abies* species (fig 12), and occasionally in Scots pine (fig 2). At its extreme it leads to mass browning and die-back as seen in figs 13 and 14 – a particularly unusual case of damage in Scots pine.

1.4.3. Wind

As already noted, the desiccating effect of wind probably plays a part in some forms of winter injury, and it also seems to provoke symptom expression in needles infected by *Lophodermium seditiosum* (see section 4.3). In general, the physiological effects of wind exposure are likely to be expressed as poor growth and "wind shaping" of crowns, and browning and die-back of windward shoots.

Physical damage by wind can produce symptoms resembling those caused by other agencies such as insects, frost, or chemicals. Such damage usually occurs early in the growing season as a result of shoots abrading against each other before they are fully hardened. In addition to outright breakage, shoots may be partially crushed or bent so that they become deformed as they harden later in the season. Abrasion injury is common in Sitka spruce and is more fully covered in section 3.3.

1.4.4. Drought

Prolonged drought may cause foliage browning and die-back. In woods and plantations damage is likely to occur first on the driest sites, for example on thin rocky soils or gravels, or on south-facing banks and knolls. Among conifers, larches (fig 15), Norway spruce, *Thuja* and *Tsuga* are particularly susceptible. In the case of Norway spruce, drought-induced decline probably forms part of the more general syndrome known as "top-dying" (section 3.4). Repeated summer drought has been proposed as one of the factors responsible for the protracted die-backs suffered by oak and ash in lowland southern Britain and in old beech on a wider scale (section 6.1). In many other species, drought that falls short of killing tissues directly may nevertheless predispose them to attack by pests and pathogens. In some broadleaves, of which beech, sycamore and birch are examples, drought stress may result in the death of discrete patches of bark, a process which is often associated with invasion by secondary pathogens and decay fungi. Such cases are difficult to diagnose because of the possibility of confusion with other diseases and disorders – such as sooty bark disease of sycamore (section 6.10) and beech bark disease (section 6.8). The situation can be made even more complicated when diseases (sooty bark disease, for example) are themselves drought-related. Predisposition to pests and diseases by drought stress is also well recorded in conifers. Increased susceptibility of pines to *Tomicus piniperda* (section 4.6.1), of larches to attack by *Ips cembrae* (section 5.4), and of various conifers to *Armillaria* root disease (section 2.2) are examples.

Direct injury by drought is uncommon except in lowland south-eastern Britain. However, drought stress as a factor in pathogenesis is probably a sporadic but nonetheless significant problem throughout eastern Britain.

Fig 15. Top die-back of young larch growing on a thin soil over rock on a steep, south-facing slope after the summer drought of 1989.

1.4.5. Lightning

Though probably not a common cause of damage in plantations, lightning deserves mention rather as a curiosity, being one of the few agents that can kill groups of large trees and leave little or no trace of its involvement. Classically, lightning kills long, more or less vertical strips of bark (fig 16), resulting in slightly curving and often intermittent scars that may extend from high in the crown to ground level. Scorching is sometimes visible on intact bark between detached strips. Large pieces of bark, or even entire limbs, are sometimes forcibly blown from trees when they are struck and there have been cases of trees being completely shattered by lightning. However, as well as causing such dramatic and unmistakeable damage, lightning can also kill trees, or parts of trees, and leave them intact and relatively unmarked. Even scars may be represented only by easily overlooked strips of dead bark. Surviving parts of affected trees may react by producing a "lightning ring" in the wood but this may be difficult to find. If trees are killed suddenly in groups, it is worth looking for scars on them and for scars or strips of dead bark on their live neighbours. A pattern composed of a central dead tree or group of dead trees surrounded by live trees with dead limbs facing inwards can also be associated with lightning but, as a diagnostic feature, it is not completely convincing without the evidence of scars. Unfortunately, the diagnosis of lightning groups in remote areas often rests on the absence of any evidence for other causes.

Fig 16. Recent lightning scar on oak; note the strip of partly detached bark.

1.5. Site

Most site-related problems involve water supply or nutrition. Drought has been dealt with under climate though, as noted there, site often has a major impact on drought damage. Waterlogging is included here because it is more frequently a site problem than a product of extreme weather conditions *per se*.

1.5.1. Waterlogging

Waterlogging frequently contributes to the failure and death of recently planted trees (section 1.1) and in that situation its occurrence and its effects are normally easy to appreciate. Elsewhere, though common, it can be rather easily overlooked as an explanation for group killing. Recent flooding should be obvious from scouring and the deposition of mud or gravel, but these signs become obscured with time, although the effects on trees might not. Waterlogging often occurs over limited areas in plantations on poorly drained soils when drains become blocked, and this may lead to the decline and death of small groups of trees – figure 17 shows an extreme case where several hectares of trees were killed. Since blockages frequently get worse over time, the damage may slowly spread, resembling the effect of a root disease. If the problem is seasonal, or has been alleviated, close examination may be necessary to spot the signs of past flooding under the ground vegetation, which often develops strongly in response to improved light levels under dead or dying trees. Roots killed by waterlogging often assume a bluish colour in the bark and outer wood. Unfortunately, this is not

Fig 17. Scots pine killed by waterlogging following failure to maintain drains. The ground conditions are indicated by the heavy growth of rushes; these trees are on the fringe of a large area of flooded ground.

a specific symptom and roots subjected to wet conditions after being killed by another agent probably develop a similar discoloration.

1.5.2. Groundwater Pollution and Toxic Soil

Groundwater pollution by road salt has already been mentioned in section 1.3.2. Other forms of poisoning are generally uncommon in forestry plantations, but are possibilities that might well be worth investigation if the affected trees are close to any kind of effluent discharge, or on a roadside, or planted on sites reclaimed from industrial use. If ground vegetation has been affected as well as trees, pollution would be strongly implicated, though in cases where flooding with polluted water has occurred it could be impossible to distinguish the effects of the chemical from those of waterlogging.

1.5.3 Lime-induced Chlorosis

Plantations on soils with a high chalk or limestone content are prone to develop lime-induced chlorosis, a disorder linked to iron or manganese deficiency. Symptoms, which often do not develop until canopy closure or later, include yellowing and thinning of foliage, premature loss of foliage, loss of apical dominance (in larches), stunting and die-back. Not only is the disorder itself potentially fatal, but declining trees are often attacked and killed by secondary pests or pathogens. Lime-induced chlorosis has been recorded in a wide variety of species but is probably best known in beech, pine and larch. However, even in these species, the disorder is not predictable as they (especially beech) can grow well in chalk or limestone areas.

1.6. Genetic Defects and Chimeras

Plantation trees occasionally exhibit abnormalities in growth or colour as a result of inherited mutation. They can also develop abnormalities during their own life span as a result of genetic aberrations in buds. This kind of defect leads to a partly deformed or discoloured individual known as a chimera. Genetic defects are not common, but those most likely to be seen in plantations are:

"fasciation" – in which branches or leading shoots become grotesquely fused and flattened;

a defect of Sitka spruce in which long snake-like branches develop with small flattened needles and no side shoots;

burr-like growths, also on Sitka spruce, in which numerous closely packed dwarf shoots develop at a point on the main stem – similar proliferation may produce a more or less spherical burr or tumour on small branches;

variegation – shoots, or parts of shoot, produce bright yellow or almost white foliage (fig 18).

Fig 18. Variegation in Sitka spruce – a chimera.

Defects like variegation are often unstable but some persist for many years, and some are permanent. The reasons for the occurrence of chimeras are not well understood. In some cases they may well arise from spontaneous mutation, but in others they might be induced by external agents, of which there are a number, including radiation, known to cause genetic damage. It has been proposed that low temperature injury to buds might be responsible for some variegations.

This kind of disorder should only affect one or two trees in a plantation and can be regarded as a curiosity – unless it can be perpetuated by cuttings and sold as an ornamental variety!

2. ROOT DISEASES OF CONIFERS

2.1. General

At present there are only three root-disease fungi that are likely to be encountered in plantation conifers in Britain: *Armillaria* (honey fungus), *Heterobasidion annosum* (Fomes) and *Rhizina undulata*. The first two are wood-rotting fungi that can live in decaying stumps or in the roots of dead trees as well as being able to attack the roots of live trees. Both *Armillaria* and *H. annosum* can kill large trees, though fatal attacks are more commonly seen on young trees; both also cause root- and butt-rot in live trees. *Rhizina* also kills sizeable trees but it is unable to cause decay. It is the most difficult of the three to diagnose.

Fig 19. Resin-bleeding at the base of Sitka spruce suffering root infection by *Armillaria* sp. [FC 51227]

Small trees infected by any of these fungi are usually killed quickly with little apparent warning, but larger trees gradually lose colour and may go through some years of reduced growth, often accompanied by heavy cone production ("stress coning"). Throughout this decline, and often before foliage symptoms become apparent, affected trees are likely to bleed abnormal amounts of resin from their lower trunks (fig 19). This is one of the most reliable indicators of a root problem, but it must be recognized that there is nevertheless a significant risk of confusing root disease in large conifers (especially spruce) with attack by bark-beetles or other phenomena that provoke resinosis. Patches of bark on the main stem from which resin bleeds should be carefully examined for insect activity before going on to consider root disease.

A fourth fungal root disease, that caused by members of the genus *Phytophthora,* merits some mention since these fungi cause serious damage to ornamentals and nursery stock in lowland Britain. Though they have not so far been associated clearly with the death of conifers in forestry plantations, there is evidence to suggest that they may cause mortality in young *Abies* species grown as Christmas trees on former agricultural ground. *Phytophthora* disease should be considered as a possibility in cases where the death of young conifers, apparently from root disease, is associated with wet ground conditions and cannot be attributed to any of the three preceding fungi, or to waterlogging. *Phytophthora* species are difficult to diagnose, even with laboratory investigation, and specialist advice is recommended if the presence of this disease is suspected.

2.2. *Armillaria* (honey fungus)

There are several species of *Armillaria* in Britain. Although all of them are pathogenic to some degree, only two, *A. mellea* and *A. ostoyae*, are likely to kill otherwise healthy trees. Of these, *A. ostoyae* is the species most commonly associated with mortality in forestry plantations, though *A. mellea* kills pines in lowland plantations in eastern England. Although the other species probably kill individual roots occasionally, and thereby enter trees to cause root- and butt-rot, little is known about their relative importance as forest pathogens in the UK. Like many other pathogens, *Armillaria* species are favoured by external stresses that weaken the resistance of the host. Consequently, the likelihood that infection will lead to serious damage or mortality is enhanced by drought or other stresses.

Armillaria species are forest fungi that inhabit stumps and the root systems of infected trees. They are able to spread through soil and litter by means of root-like structures (rhizomorphs) which have been given the common name of "bootlaces", though it is worth commenting that in many situations rhizomorphs are considerably finer and more fragile than this name implies. Growth as "mycelial sheets" (see later) through infected root systems is probably also an important means of spread for the highly pathogenic *A. ostoyae* and *A. mellea*. Trees are often killed by infections near the root collar and, thereafter, the rest of the dying root system can be rapidly invaded, providing a network of easy pathways through the soil for the fungus. The less pathogenic species, which have less opportunity to spread in this way through killed roots, tend to be the more prolific producers of rhizomorphs. The extensive rhizomorph networks formed by these species make them well placed for rapid colonization of root systems killed by other agents. A high density of rhizomorphs does not, therefore, necessarily mean a high risk of disease. Nor does an apparent absence or scarcity of rhizomorphs indicate low risk – severe infection may occur on sites where rhizomorphs are extremely difficult to find.

Rhizomorphs serve as a means of infection as well as dispersal but, once they have penetrated the bark, spread of the fungus within the living host is achieved by another type of fungal aggregation called "mycelial sheet" (fig 20). This is a coherent, white or creamy skin of about paper thickness or slightly thicker. It is quite fragile and, from its method of growth, sometimes has fan-shaped lobes at its edge. The presence of mycelial sheets in the bark or between the bark and wood is the most reliable indicator of attack on a living tree by *Armillaria*. Although confirming the presence of *Armillaria*, contrary to popular belief, rhizomorphs are neither a reliable diagnostic feature nor a straightforward guide to site hazard.

Fig 20. Mycelial sheet below the bark at the base of a young Sitka spruce. This is the most reliable indication of *Armillaria* killing. [Les Starling].

In contrast to the ease with which *Armillaria* species can travel below ground, long distance spread from site to site by means of spores is much less efficient. Thus, *Armillaria* disease is only likely to be a significant problem on sites with a long woodland history, broadleaf or conifer, where it is already established. It would be unusual for long-standing woodland not to have one or more species of *Armillaria*, but these need not necessarily be highly pathogenic. The frequently repeated view that *Armillaria* disease is confined to old broadleaved sites is a myth which has arisen from the particular history of forestry in this country.

There is little doubt that *Armillaria* will gradually spread into our new upland forests, but high mortality in first or second rotation stands on previously tree-less hill ground is unlikely to be due to it. Significant losses can occur, however, in the second rotation of conifers on ground that originally carried scattered scrub.

Armillaria killing is usually first noticed on restocked sites three to five years after planting, when scattered individuals turn yellow and die. Death is often quite rapid in young trees. As the disease progresses, groups of dead trees tend to form and may eventually leave large gaps in the crop. Mortality can probably continue throughout the life of a crop (there are records of killing in Sitka spruce and Scots pine over 50 years old), but it normally declines after 15 years or so as resistance increases with age. Most forest hardwoods are resistant – even those such as *Salix*, *Acer* and *Prunus* that are recorded as susceptible are likely to be more resistant than conifers. Oak is extremely resistant. Most commercially grown conifers are susceptible, but noble fir, grand fir, Douglas fir and larch are reputed to have some resistance to killing. These species may nevertheless suffer root-rot, which could reduce stability later in the life of the crop.

As a group, *Armillaria* species are some of the easiest fungi to identify if rhizomorphs are present, yet their status as pathogens can be extraordinarily difficult to judge. All the British species, irrespective of their pathogenic ability, are effective secondary colonists of trees killed by other agents and so, on sites where *Armillaria* is present, it is inevitable that a proportion of trees killed by any agent will be invaded. The killing of large conifers, while far from unknown, is sufficiently uncommon that apparent attacks on trees over 30 years old should attract careful investigation for other agents that might bear primary responsibility. If broadleaves are involved, even more caution would be necessary.

The most effective control measure against *Armillaria* disease is to remove the stumps that harbour the fungus. While the expense of such an operation would be unjustified on most forest sites in Britain, it might be seriously considered if losses were to become severe on sites with light soils and good access for machinery. Where destumping is impractical, losses can be reduced by using resistant species, but, on severely affected sites and in gaps that develop as a result of attack in young crops, hardwoods may be the only option. Large stumps containing the fungus remain a source of infection for decades.

2.3. ***Heterobasidion annosum*** **(Fomes)**

In contrast to *Armillaria*, spores of *H. annosum* readily colonize freshly cut stump surfaces, enabling it to spread over long distances between forests and to build up rapidly within forests. Like *Armillaria*, it lives by decaying wood and has no independent existence in the soil. Consequently, unthinned first rotation crops are extremely unlikely to harbour the disease and may be maintained free of it by treating freshly cut stumps to prevent infection by airborne spores (see later). As this control measure was introduced in 1960, infection is most likely to be encountered on sites which were woodland in the 1930s and earlier. The highest levels of disease are usually found on sites with a woodland history extending for two or more rotations.

Unlike *Armillaria*, *H. annosum* has no means of spreading freely through the soil. It is confined to root systems and, in consequence, infection only occurs where the roots of healthy trees are in contact with roots or stumps of infected trees. Thus, unless new foci of infection are created by thinning, the disease spreads relatively slowly.

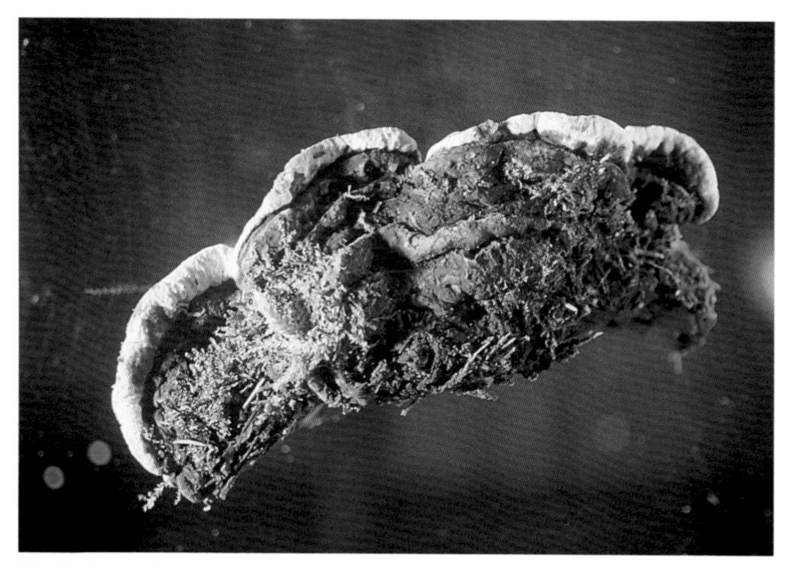

Fig 21. Fruit body of *Heterobasidion annosum*; a bracket viewed from above.

The presence of *H. annosum* on a site may be revealed by fruit bodies on stumps or on dead trees (figs 21, 22 and 23) but, even in these cases, there are often few indicators of the severity of disease. Site history and soil type are the most reliable guides to the likelihood of infection.

H. annosum causes root- and butt-rot in most commercial conifers and, on site types that favour the disease, young trees of most species may be killed. High risk sites are primarily those with mineral soils, especially ex-arable sites; soils with high pH present the greatest risks. In contrast to other commonly planted conifers, pines can be killed at all ages by *H. annosum* and on some sandy, high pH soils mortality may be so severe that the removal of infection sources by destumping between rotations is necessary to produce acceptable crops. Because

Fig 22. Fruit body of *Heterobasidion annosum* showing the creamy white pore surface underneath. The bracket is attached to an exposed root.

Fig 23. Fruit bodies of *Heterobasidion annosum* on a dead lodgepole pine: at the ground line close to the side branch is a typical horizontal bracket partly covered by needle litter; extending up the main stem are smaller, immature, button-like fruit bodies that are often found on infected trees.

the original research on *H. annosum* in Britain was carried out on such sites in eastern England, experience there has tended to colour perceptions of the disease elsewhere, and some misunderstandings have gained wide currency as a result. Although in eastern England much higher mortality among pines occurs on sands with a pH above 6 than on more acidic soils, this observation cannot be extended to predict a universally low risk of decay in other species growing on acidic soils in the uplands.

In species other than pine, *H. annosum* is more important as a cause of butt-rot than mortality and, indeed, can be regarded almost as a different disease. The appearance of the rot is shown in fig 69 and briefly described in section 8.3. Without stump protection, losses from butt-rot in second or third rotation spruce on mineral soils (including those with pH below 5) may reach 30% of volume, with 70–80% of trees infected. Losses on peaty soils, by contrast, may be negligible even without stump treatment. The analysis of risk, and hence the expected benefit from stump treatment, is a complicated process in upland forests and specialist advice should be sought.

The recommended material for stump treatment is currently a 20% solution of urea applied to the point of run-off immediately after felling. However, the efficacy of alternative chemicals, such as disodium octoborate ("polybor"), is currently being investigated. Although chemical stump protectants are effective on pines, the biological control agent *Phlebiopsis (Peniophora) gigantea* is equally effective and has a number of advantages.

Stump treatment is unlikely to have a significant effect in reducing disease on sites where *H. annosum* is already well established. Under these circumstances, control can only be achieved in future rotations by stump removal or, where this is impractical, by the use of resistant species such as broadleaves. Some conifers, notably grand and noble firs and, to a lesser extent, Douglas fir, are relatively resistant to butt-rot, though some killing can be expected in young noble and Douglas fir on infected sites. Douglas fir may also suffer root infection which might increase the risk of windthrow. It should be emphasized that the relative resistance of grand and noble firs may not be maintained in old trees, or on all sites, nor does it extend to all *Abies* species – *A. amabilis*, a species sometimes advocated for wider use in UK forestry, is extremely susceptible to *H. annosum*.

2.4. *Rhizina undulata*

In cases where a root disease is indicated, but no evidence of *H. annosum* or *Armillaria* can be found, *Rhizina* should be considered as a possible cause, especially in pole-stage crops. It is strongly associated with fires and is the most likely explanation of group killing if there are traces of fire inside groups. The presence of *Rhizina* would be confirmed by finding fruit bodies on the ground in summer, but their occurrence is not always to be relied on. They are dark red or chestnut brown, irregularly rounded, lumpy structures with a distinct cream-coloured margin when they are growing actively (fig 24). Despite their appearance, the fruit bodies are hollow underneath, and their lower surface is attached to the

Fig 24. *Rhizina undulata* fruit bodies in July (viewed from above) growing on the site of a fire in the spring of the previous year.

ground by yellowish white strands. Fruit bodies can reach about 10 cm in diameter and may appear inside or around the edges of the group of trees killed by the fungus.

Rhizina spores, which are widely dispersed in forest soils, are activated by the high temperatures (in the order of 30–40°C) likely to be reached in the ground under fires. Having started growth, the fungus spreads through the soil and attacks fresh conifer roots, including those of recently cut stumps. Once established, it continues to spread and causes an expanding infection centre. *Rhizina* grows through the soil as yellowish white threads known as mycelial strands. Although less easy to see and identify than *Armillaria* rhizomorphs, they can aid diagnosis. They ramify over the surface of roots, attacking the bark at many points and causing discrete resinous lesions. Roots are gradually overwhelmed and killed, and eventually trees die in an expanding group. Mycelial strands and root lesions are most likely to be seen on trees under attack at the edges of groups. Conifers of any age may be attacked, but the disease is most common in pole-stage stands, and Sitka spruce is particularly susceptible. Groups normally continue to spread for between five and ten years, after which the fungus ceases activity, though losses from wind-throw of trees with damaged root systems may continue. In the later stages of spread the fungus appears to remain active only at the margin, and this may be the only area where fruit bodies are produced – if they are produced at all.

Rhizina can cause loss of newly planted trees on sites cleared by burning lop and top before restocking. Under these circumstances the fungus does not appear capable of spreading far and so, if the fires are confined, losses are unlikely to be severe. Broadcast burning, by contrast, has the potential to precipitate high mortality in the succeeding crop if high soil temperatures occur on a wide scale. In practice, however, this type of fire has rarely been associated with restocking losses in upland plantations in the UK.

The association of *Rhizina* with bonfires is very strong, but not absolute. There have been a few cases where the fungus has almost certainly been responsible for group killing in spruce crops but no traces of fire could be found. The initiation of these groups remains a mystery.

3. DISEASES, DISORDERS AND PESTS OF SPRUCE

Although the two spruces commonly grown in UK plantations, Norway spruce and Sitka spruce, share a number of diseases and disorders, they also display some differences in pathology. For example, Norway spruce suffers badly throughout Britain from a fatal disorder known as top-dying which is unknown in Sitka spruce. On Sitka spruce, by contrast, the commonest problems seen in forestry plantations are probably frost injury and aphid attack, both of which trouble Norway spruce much less.

3.1. Frost

Although frost rarely causes economic loss, its potential for damage to Sitka spruce places it alongside *Elatobium* and decay as one of the most serious health problems that can affect this species. Frost damage is a major consideration in the choice of provenance and, over most of northern Britain, the benefits of faster growth by southerly provenances have to be foregone because of their corresponding vulnerability to frost. The long growth period of more southerly provenances puts unhardened tissues at risk from the early season and late season frosts that are a feature of the British climate. Frost damage is a significant factor only in small trees and is rarely fatal. Nevertheless, the killing of buds and shoot tips in spring or autumn, and of new shoots in spring, can hold back height growth, or even cause complete check on the most frost-prone sites. Recovery growth in Sitka spruce is usually vigorous, but form may suffer and rotation length may be extended, both of which may incur financial penalties. Norway spruce is much less susceptible to frost injury since it has a shorter growing season than most of the commonly planted provenances of Sitka spruce.

Fig 25. Spring frost damage to newly flushed Sitka spruce shoots.

Diagnosis of frost damage is not without pitfalls. Although "tasselling" of newly flushed Sitka spruce shoots at the start of the season (fig 25) is most likely to be spring frost injury, such damage is nevertheless worth a closer look in young plantations because of the possibility of confusion with winter moth damage. The effects of frost can also be confused with those of two other insects, *Elatobium* and *Zeiraphera ratzeburgiana*. All three are dealt with in the following two sections.

Fig 26. Twisted one-year-old shoots on Sitka spruce resulting from non-lethal frost injury in spring or early summer of the previous year.

Spring frost injury causes current shoots, especially on lower branches, to turn yellow or light brown, wither and, if they have reached any length, hang limply down. On small trees, all shoots may be affected (fig 25). Later, the damaged tissues darken and, if the shoots had barely emerged from the bud when they were frosted, they may persist for a year or more as dark brown or black tufts. In contrast to insect-damaged shoots, frosted shoots remain intact and recognizable as complete (albeit undeveloped) shoots without mining or frass among the needles. All affected shoots tend to be about the same size, consistent with a single incident rather than a developing infestation. Foliage of the previous year usually remains undamaged.

Non-lethal frosts, which usually occur in early summer, can cause still-soft current shoots to become flaccid. This may lead to permanent deformities if the shoots continue to grow while in this state (fig 26). Non-lethal frosts can also cause needle yellowing on current shoots.

Fig 27. Pre-flushing frost injury on Sitka spruce, of Alaskan origin, caused by frosts in April; photographed in the following September. Despite having shown more or less normal extension growth, the discoloured top is dying as a consequence of bark injury (on the main stem at the position marked by the red tape) which occurred before flushing.

A particular type of damage may follow late April or early May (i.e. before flushing) frosts if they are preceded by a spell of unusually warm weather. The warm weather induces cambial growth, particularly on the main stem, so that large patches of bark can be killed by the subsequent low temperatures. This results in die-back of branches or entire tops of trees up to several metres in height (figs 27 and 28). Die-back is sometimes associated with lesions that are visible (but not obvious) as depressed, cracked or resinous areas of bark. It is a feature of this kind of frost injury that affected trees are able to flush and grow almost normally for weeks or months before the full effects become apparent, as in the case illustrated in fig 27. "Pre-flushing frost damage" is not common but the symptoms associated with it are unusual and may be rather alarming. Typically, scattered individuals or small clusters of trees are affected, but damage can be more extensive (fig 28).

Damage by autumn frost is common in Sitka spruce in northern Britain. In September or October night temperatures of about –5°C (grass minimum), or lower, that follow mild late summer or early autumn weather can cause particularly distinctive symptoms: current needles turn brown while older needles, if they are present, remain green (fig 29). The least affected needles develop a central, pinky brown to red–brown band, in which the pink hue is often most strongly developed on the underside of the needle. The worst affected needles become completely reddish or pinky brown with, again, any pinkness being more evident on the lower surface. Although all the current needles may be affected, on some shoots needles at the shoot tip are likely to remain green (fig 29).

Fig 28. Pre-flushing frost injury at the same site and at the same time as the damage shown in fig 27, but on a more southerly provenance of Sitka spruce (Vancouver Island).

This type of autumn frosting could be confused with autumn attacks by *Elatobium* (next section) or, conceivably, with damage by the conifer spinning mite. The latter tends to develop outwards from the centre of the crown, although all foliage may eventually be affected. Attacked needles are at first speckled dull yellow, then take on a bronzed or coppery appearance. Serious infestations by spinning mite are associated with silky webbing, but this may be sparse and difficult to see.

Fig 29. Autumn frost injury on Sitka spruce needles; only current foliage is affected.

Autumn frost damage can occur spectacularly over large areas, but the consequences are generally not serious because injury is usually confined to needles. Damage to shoots and buds is likely only after exceptionally severe frosts, or on very frosty sites, or if southerly provenances are used on frost-prone sites. Under such circumstances unhardened shoot tips and buds, especially the late season ("lammas") growth of young Sitka spruce, are killed (fig 30). By the following season, the damage is likely to be distinguishable from severe spring frosting only by the state and size of the dead buds and the presence or absence of tasselled new shoots.

Fig 30. Killing of shoot tip and buds of Sitka spruce by autumn frost; the shoot has been cut open to show the distribution of dead tissue.

3.2. *Elatobium abietinum* (Green Spruce Aphid)

The following section refers principally to Sitka spruce on which heavy defoliation by *Elatobium* is common and likely to lead to reduced growth.

Normally, *Elatobium* attacks mature needles in winter, spring and early summer. Thus it mostly affects needles of the previous year – or the current year in the case of damage occurring early in winter. Attacked needles firstly develop yellow or pale brown bands, but eventually discolour completely, though they often retain a mottled look, and fall. The presence of the aphids themselves (they are bright green, just visible to the naked eye, and have distinctive red eyes if seen through a lens), or their cast skins on attacked needles, is diagnostic. Unfortunately, live insects have often disappeared by the time symptoms are seen. The only remaining evidence may be a sticky, sugary deposit ("honeydew") that coats needles and shoots and eventually turns black as it is colonized by fungi ("sooty moulds"). On larger trees bordering roads or other gaps, damage is usually most severe in the lower crowns, but in exceptional years it may extend into the upper crowns of interior trees to cause a noticeable increase in the light level inside stands.

Only in the very heaviest attacks, when all the old foliage may be shed, do the insects feed on needles of the current season, doing so on the oldest needles at the base of the new shoots. In attacks of this severity, the newly flushing shoots may develop longitudinal stripes of yellow or straw-coloured needles, or may wither completely when only a few centimetres long. Such damage looks remarkably like spring frosting but can usually be recognized by its association with signs of heavy *Elatobium* attack on the needles of the previous year, and by the occurrence of a proportion of yellowed but unwithered shoots. This type of damage may be found in young (pre-thicket or early thicket) trees or, most characteristically, on semi-suppressed shoots and epicormics in the lower crown of large roadside or rideside trees. It is possible to mistake such damage in young trees for attacks by winter moth, but needles on *Elatobium*-affected shoots remain intact (though they may fall) and there is no sign of mining or frass inside the withered shoots.

On rare occasions, *Elatobium* attacks develop in autumn on mature needles of the current season. In such cases, affected needles tend to be retained long enough for the browning to resemble autumn frost damage. However, the usual signs of the insect should be evident if the damage is seen early, and the browning will lack the regular patterns described in the preceding section on frost.

Other than in Christmas tree plantations, *Elatobium* damage on Norway spruce is less common than on Sitka spruce and normally much less severe. Symptoms on Norway spruce are similar to those described above, but the initial yellow and brown banding is less noticeable and completely brown needles can be retained on affected trees for some time.

3.3. Other Causes of Injury to Young Sitka Spruce Shoots

Feeding by winter moth (*Opheroptera brumata*) can kill young developing shoots which then wither to give a tasselled appearance very like frosting. However, if winter moth is to blame, the shoots will show signs of physical damage and frass may be visible. Moreover, injury is unlikely to be confined to the destruction of small shoots. Larger shoots, dead or alive, should also be present with the mixture of symptoms characteristic of winter moth: missing and partially eaten needles, resinous scabs and lesions, and deformation. This type of damage is in marked contrast to the uniform tasselling or drooping of frosted shoots.

Another insect, the spruce tip tortrix (*Zeiraphera ratzeburgiana*), mines flushing buds during spring to cause a range of symptoms including abortion of the whole flush, which again superficially resembles spring frost injury. As with winter moth, frass should be evident inside the outer needles of the dead shoot.

Abrasion of shoots against each other by wind in early summer causes symptoms almost identical to those of winter moth, but without frass. However, it should be emphasized that frass is not always easy to find in genuine winter moth attacks. Abraded needles tend to be shrunken and to have puncture wounds rather than signs of chewing, though microscopic examination may be necessary to make sure of this. On some shoots there will be patches that are rubbed smooth rather than scarred, and damage is likely to involve older shoots, which winter moth does not attack. Wind may also subject soft young shoots to bending and twisting that create permanent deformities. When shoots are examined *in situ*, a diagnosis of wind damage can usually be confirmed by pushing adjacent branches together to simulate wind action and by matching adjacent abraded shoots. Common sense often rules out wind abrasion as a likely cause, but Sitka spruce in upland plantations thrashes about to an extraordinary degree in high winds and abrasion is a constantly overlooked cause of deformation and damage to young shoots. While far less damaging than an outbreak of winter moth, abrasion damage to leading shoots on young trees could lead to the eventual deformation of lower stems. Gales early in the growing season often cause leaders to break at or near the junction with the previous year's growth. The level of damage varies with exposure and elevation but can be serious, especially in fast growing crops.

3.4. Top-dying of Norway Spruce

This disorder is a common and frequently significant cause of decline and death in Norway spruce, especially on the eastern side of Britain. It usually affects trees from pole-stage onwards and, with the exception of rare cases on *Picea omorika* and *P. glauca*, it is confined to Norway spruce. Foliage on affected trees turns brown, usually from the shoot tip backwards (fig 31), and usually in late winter or spring, though it may happen much earlier. The development of browning is variable – it may occur simultaneously over a large part of the crown, or on branches scattered throughout the crown, or on the windward side of the crown. Trees can recover from a single episode of foliage browning, but in many cases it recurs so that crowns thin and then progressively die back. Top-dying is nearly always accompanied by a fall-off in height increment, the onset of which may precede foliage browning.

Typically, top-dying only affects a proportion of trees in a crop so that healthy, browning and dying trees may stand side by side (fig 32). Another important feature of the disorder is that on trees with advanced crown symptoms, the roots and lower stem remain alive. This makes a useful distinction from root diseases and disorders, which would be the most likely explanation, apart from top-dying, for the death of large Norway spruce.

Fig 31. Discoloration of needles at the shoot tips of Norway spruce affected by top-dying.

The cause, or causes, of top-dying are not known. There is no evidence that living agents are involved – indeed, several features of the syndrome strongly suggest that it is abiotic in origin. The most convincing hypothesis relates it to water stress induced when transpiration exceeds water uptake. It can, therefore, be precipitated by excessive water loss, or by a restricted supply, or by a combination of both. Poor ability to control water loss may be a feature of a continental species such as Norway spruce grown as an exotic in a maritime climate. This explanation accounts for the strong associations of top-dying with stand edges, exposure to air movement, drought, and mild, windy winters. The last are believed to allow transpiration to continue while roots are relatively inactive. Clearly, this disorder verges on drought injury and in times of severe summer drought, as in south-eastern England in recent years, it becomes academic to maintain the distinction.

Top-dying is usually concentrated on crop edges, though affected individuals and small groups may also be scattered throughout stands, especially those on steep slopes. Trees around gaps inside larger stands are also vulnerable, and the disorder can develop in crops that are opened up by thinning. Delayed heavy thinning is

Fig 32. Top-dying of Norway spruce.

particularly dangerous. Top-dying sometimes becomes a severe problem in small isolated stands where all trees are close to an open edge, but it probably reaches its worst development on new edges created by adjacent clear-felling. In northern Britain, its occurrence is virtually inevitable if such edges face west or south-west. Small remnant blocks of Norway spruce left in clear-felled coupes are at considerable risk of rapid decline.

The condition may be precipitated several times in the life of a crop and it varies in its development on individuals. Consequently, affected stands often have a mixture of long-dead trees, recently killed trees, browning trees, more or less green trees with reduced height increment, and healthy trees.

The management of crops with top-dying can be difficult, but the gradual removal of affected or dead trees during normal thinning is probably the best course of action. Heavy thinning should be avoided because it is likely to exacerbate the problem. Similarly, the removal of an unsightly concentration of damaged trees on a stand edge will create another new edge along which top-dying will probably develop. In small stands losses may become high enough to warrant clear-felling.

3.5. Nutrient Deficiencies

There are three elements that are commonly deficient in plantation spruce: nitrogen, phosphorus and potassium. The following refers to symptoms on Sitka spruce, in which they are best known.

Potassium deficiency produces a distinctive yellowing of older foliage (fig 33). This is visible throughout the growing season but is particularly evident late in the season, when current needles may also start to yellow in severe cases. At its most developed, the discoloration is a hard, dull yellow that generally affects needles uniformly, but may be more accentuated in the centre. Discoloration, even when severe, may affect only occasional individuals and may also only develop in patches of the crown. It is a photo-effect so that shaded branches directly underneath severely yellowed ones may be relatively normal in colour. Chronic and severe potassium deficiency can lead to die-back that is characteristically confined to subsidiary shoots while the main axis of the affected branch grows on. Potassium deficiency is only likely to develop to any extent on peaty soils.

Fig 33. Yellowing of older needles in Sitka spruce due to potassium deficiency. Note that some subsidiary pendant shoots have defoliated tips and have died back (or are inactive). On all active shoots, the current foliage is a normal blue-green colour.

Nitrogen-deficient trees develop a bright yellow-green colour which, in contrast to potassium deficiency, is strongest in the youngest needles. Again unlike potassium deficiency, nitrogen deficiency tends to cause shortening of needles. Severe and chronic nitrogen deficiency, which is most likely to be seen where spruce is growing in competition with heather, leads to declining growth and crop failure ("heather check").

Phosphorus deficiency has relatively little effect on colour, but needle length is reduced, ultimately to scale-like proportions when the deficiency is severe. Shoot length is also reduced and die-back occurs in extreme cases.

3.6. *Chrysomyxa* Rusts

There are two species of *Chrysomyxa* that attack spruce in Britain: *C. abietis* and *C. rhododendri*. They occur on both Norway spruce and Sitka spruce, but *C. abietis* is more common than *C. rhododendri* and is rare on Sitka spruce. Though symptoms on individual trees can be dramatic, serious damage by either rust is unlikely since infection is sporadic and uncommon. From a distance, infected trees look yellow, resembling severe nutrient deficiency, but close examination should reveal that, on some needles at least, the yellowing is confined to sharply delineated bands or blotches (figs 34 and 36). *C. abietis* infects young needles in early summer. They become mottled with yellow blotches throughout the late summer and autumn, and are bright orange-brown by the end of winter. Pustules (fig 36) are produced in early summer and release spores to infect the new flush of needles. If seen only in winter or early spring, when the fungus has not produced its distinctive pustules, this disease could be confused with the effects of frosting or *Elatobium* attack from the previous autumn. *C. rhododendri* also infects in early summer but it develops pustules by mid or late summer (figs 34 and 35) to infect rhododendrons, the alternate host in which this fungus overwinters. The yellow blotches or overall bright orange-brown colour in summer, affecting only the current foliage, are unlikely to be mistaken for anything else.

Fig 34. *Chrysomyxa rhododendri* on Sitka spruce; symptoms on current needles in late September. Note that the pustules have not yet opened to release spores as in fig 35.

Fig 35. *Chrysomyxa rhododendri* on Sitka spruce; symptoms on current needles in late September, detail of sporing pustules.

Fig 36. *Chrysomyxa abietis* on Norway spruce; symptoms on needles of the previous year in May.

3.7. *Cucurbitaria* Bud Blight

Since this disease tends to occur high in the crown of sizeable trees, it often escapes notice. It is probably quite common on old Norway spruce but, with the exception of a single doubtful record from the 1930s, it is not known on Sitka spruce. The disease causes swelling of buds and some distortion of shoot tips; infected buds usually die, but may produce feeble, short-lived shoots which are often hooked back. Infected branch systems take on an irregular and stunted appearance with a zig-zag branching pattern and with the swollen buds giving a clubbed look to individual shoots. Because of the loss of buds in affected crowns, the distorted branches are often seen in an otherwise thin, spire-like top. This collection of symptoms can be recognized in the crown (fig 37) with binoculars, though confirmation of the presence of the disease requires close examination of detached shoots. The causal fungus, *Cucurbitaria piceae*, characteristically produces a black encrustation on infected buds.

Fig 37. Crown of mature Norway spruce showing irregular branching caused by infection of buds by *Cucurbitaria piceae*. [FC 51222]

4. DISEASES, DISORDERS AND PESTS OF PINES

4.1. Potassium deficiency

This causes foliage yellowing which is particularly marked at the needle tips. In most cases on lodgepole pine, yellowing affects only the older needles, but in Inland provenances, and in Scots pine, it may extend to current foliage. In cases of severe deficiency, needles may develop necrotic tips. The yellow appearance of potassium-deficient lodgepole pine crops is usually unmistakeable, but it does vary in pattern and intensity and, at a casual glance, some extreme cases could be confused with *Ramichloridium* infection (section 4.6.3). As with spruces, potassium deficiency in pines is only likely to develop to any great extent on peaty soils.

It is worth noting that on non-potassium-deficient sites individual lodgepole pine are occasionally seen with dramatic late-season yellowing of current foliage. The reason for this is not known.

4.2 *Coleosporium tussilaginis*

C. tussilaginis is a rust fungus that spends part of its life cycle on pines (usually Scots pine, occasionally Corsican pine) and the other on a range of herbaceous plants including *Senecio* (groundsel and ragwort), *Tussilago* (coltsfoot), *Sonchus* (sow-thistle), *Euphrasia* (eyebright) and *Campanula*. Symptoms on pine appear at any time from March onwards on needles of the previous year only. The first signs are small yellow spots that sometimes ooze sticky droplets, but these are soon followed by the production of prominent white sacs containing orange spores. These distinctive fruit bodies form on yellow blotches while the rest of the needle often remains green. Even after the spores have been dispersed, the ragged white skins of the burst sacs may remain attached to the needle.

C. tussilaginis is likely to cause significant damage only in nurseries and Christmas tree plantations, where appearance is important.

4.3. *Lophodermium seditiosum*

This is the most common and serious foliage disease of Scots pine. In some years, current needles turn a mottled yellow as early as autumn, but the disease in not usually noticed until late winter or spring when infected parts of the (by then) previous year's needles turn blotchy, reddish or purplish brown. Discoloration may affect the entire needle or only part of it; the presence of partially affected needles with green tips is the best indication that browning is not abiotic in origin, but examples can be difficult to find by the time most attacks are seen. *L. seditiosum* may affect part of a shoot or cause complete browning and defoliation. It is usual for damage to be worse in the lower crown, but thicket to pole-stage trees can be entirely defoliated. On open sites, damage often develops more severely on the windward side and has frequently been mistaken for exposure or "winter blast".

The symptoms caused by *L. seditiosum* could also be confused with those of *Lophodermella sulcigena*, which is another fairly common needle disease of Scots pine and which is dealt with in the next section.

Fig 38. Scots pine plantation affected by *Lophodermium* needle-cast.

Fig 39. Scots pine showing severely reduced increment as a result of defoliation by *Lophodermium seditiosum*. Photographed in May 1987; the bottle-brush-like shoots are those formed in 1986 following complete loss of the 1985 (and older) foliage.

Infection by *L. seditiosum* occurs from mid-summer to autumn on current needles by means of spores released from fruit bodies on the old infected needles, most of which have fallen by then. Severe damage in spring is therefore largely a reflection of suitable conditions for the fungus in the previous summer or autumn. The effects in spring may be dramatic, with entire plantations turning reddish brown (fig 38), apparently quite suddenly.

Shoot extension may be sharply reduced in the season following defoliation so that the upper whorls assume a "bottle-brush" appearance (fig 39). *L. seditiosum* does not attack shoots directly, but the indirect effects of severe foliage damage can, in exceptional years, lead to some shoot mortality. However, care is required in attributing die-back to *L. seditiosum* as coincident attacks by it and the shoot pathogen *Brunchorstia pinea* are not uncommon (section 4.6.4).

4.4. *Lophodermella sulcigena*

This disease is common on Corsican pine of all ages and on young Scots pine. On Corsican pine it is generally most noticeable in late winter or early spring as a grey-brown discoloration of the previous year's needles (fig 40). Characteristically, the base of infected needles remains green. The disease may affect only scattered needles but attacks are often severe enough to impart a greyish brown appearance to entire trees or stands. Such attacks may lead to growth reductions.

Fig 40. *Lophodermella sulcigena* on Corsican pine; symptoms in February.

On Scots pine, the disease becomes evident in late summer or autumn as a purplish or pinkish brown discoloration of current needles (fig 41). On badly discoloured needles, green bases or tips (or both) should be apparent with a sharp division between green and discoloured tissue. Infected needles eventually become entirely greyish brown though they often keep a mottled appearance. It is not unusual for infection to be confined to one needle of a pair. In contrast to the normal situation in infected Corsican pine stands, the disease does not commonly affect more than a few individuals or small groups of trees in Scots pine stands. Infected trees may show quite severe needle browning while neighbouring trees remain completely unaffected. Badly infected individuals can present quite a striking purplish appearance.

Fig 41. *Lophodermella sulcigena* on Scots pine; symptoms in early October.

Browning seen in spring on Scots pine could be due to infection by either *L. sulcigena* or *Lophodermium seditiosum*. At this stage it may be impossible to distinguish between them without laboratory examination – and difficult even then. However, by spring needles infected by *Lophodermella* are likely to look greyish, whereas the discoloration produced by *Lophodermium* nearly always has an element of reddening. Extensive browning of any hue in a Scots pine crop is unlikely to be due to *Lophodermella*.

4.5. *Peridermium* Canker

The stem-rust fungus *Peridermium pini*, which is common only in north-eastern Scotland and East Anglia, attacks Scots pine and, rarely, Corsican pine. It is a bark pathogen that infects young shoots to cause small, swollen cankers, though these may be barely noticeable in the early stages of infection. They should not be confused with the very inflated, ovoid, woody swellings that are sometimes found on young shoots and that are induced by mites. *Peridermium* infection gradually spreads down the shoot and branch, extending the canker towards the main stem. The distal part of the branch stays alive for some years and when it dies it often remains recognizably swollen and covered in a blackened exudate of resin. In May or June the fungus may produce unmistakable pale yellow or orange pustules, about 5 mm across (fig 42). As these occur only on live bark, they are typically produced around the margin of large cankers. Old pustules leave crater-like marks which, when they are numerous, impart a roughened appearance to the dead bark of older cankers. If the fungus reaches the main stem, it can girdle it and so kill the tree above that point (fig 43) – cankers below the live crown can kill the tree completely. Main stem cankers are extremely resinous and, when old, usually exhibit a central blackened area of dead, cracked and resinous bark from which pieces may fall to leave areas of exposed wood. Old cankers that have almost girdled the stem are often made noticeable by a rope-like swelling along their length resulting from callus growth produced by the narrow, surviving part of the stem circumference.

Fig 42. *Peridermium pini* on Scots pine; pustules on a branch canker.

When they occur on the main stem of Scots pine, *Crumenulopsis* cankers (section 4.6.6) can be mistaken for *Peridermium* cankers. The former usually have a less severe effect on infected stems than *Peridermium*, and are less common than *Peridermium* in north-eastern Scotland and East Anglia, but it requires considerable experience to be sure of the distinction between these two pathogens.

Fig 43. *Peridermium pini* on Scots pine; dying top resulting from a girdling canker which can be seen as a blackened area on the main stem immediately above the upper whorl of live branches.

4.6. Shoot Die-back

There are six common causes of die-back in young (1 to 3-year-old) pine shoots:

> *Tomicus piniperda* (pine-shoot beetle) – on Scots, Corsican and lodgepole pine;
>
> *Crumenulopsis sororia* – on Scots, Corsican and lodgepole pine;
>
> *Brunchorstia pinea* – on Scots and Corsican pine; very rare on lodgepole pine in UK;
>
> *Ramichloridium pini* – on lodgepole pine only;
>
> winter desiccation – common on lodgepole pine, occasional on Scots pine;
>
> frost – likely only on young Corsican pine.

With the exception of *Tomicus*, these can be difficult to distinguish on the basis of field symptoms.

4.6.1. *Tomicus piniperda* (Pine-shoot Beetle)

During maturation feeding in mid to late summer, these insects bore up the centre of current-year shoots. The resulting dead, hollow shoots are easily recognized if they are seen at close quarters. In attacked crops, dead or dying shoots are evident at most times of the year, but they finally break off – in fact, the diagnosis of *Tomicus* in a stand often rests on the presence of broken, hollowed shoots on the ground. Shoot killing by *Tomicus* may reach significant levels in pines (Scots, Corsican or lodgepole) growing close to pine logs in which breeding has been allowed to take place. Although most breeding is done in the bark of newly dead or felled trees, *Tomicus* will attack the main stems of live trees for the same purpose. Such attacks are normally unsuccessful, but if the trees are under stress they can be overwhelmed and killed. Apart from drought, the commonest situation in which fatal attacks occur on standing trees is that of crops already attacked by *Lophodermium* or *Brunchorstia* (section 4.6.5).

4.6.2. Winter Injury

Winter injury commonly affects one-year lodgepole pine shoots on exposed sites in northern Britain. Though the following description is based principally on observations of lodgepole pine, Scots pine occasionally suffers similar damage. In late winter or early spring, needles on affected shoots turn dull green (i.e. a loss of the normal bright, shiny tone) and progress through olive green, bronze and tan to bright red-brown (fig 44). Affected shoots normally fail to flush and eventually die back. However, it is typical of winter injury that,

Fig 44. Winter injury to lodgepole pine shoots. Photographed in July; current growth can be seen on an unaffected shoot in the background.

at the time of flushing, buds may fail to extend on some slightly discoloured shoots although the bark, and the buds themselves, are alive. This is one distinction from the common fungal diseases of shoots since any effect the latter may have on bud extension is likely to be associated with bark necrosis. Other important diagnostic features of winter desiccation injury relate to the distribution of damaged tissues. Most shoots with a significant proportion of symptomatic needles eventually die back, but the dead bark does not necessarily extend down the full length occupied by discoloured needles. Moreover, a few discoloured shoots may survive. In the case of die-back caused by *Ramichloridium* or *Brunchorstia* (next sections), which are bark pathogens, death and discoloration of needles is consequent on the killing of bark. Therefore, the zone of dead bark either coincides with or extends beyond that of dead needles (fig 48). The distribution of needles and shoots affected by winter desiccation is sometimes clearly related to wind direction and, in lodgepole pine, this can be a valuable pointer to the cause of damage. However, it is not so reliable in Scots pine (since *Lophodermium* needle-cast can be influenced by wind exposure) and, even in lodgepole pine, there is frequently no convincing pattern of injury. Symptoms of winter injury develop rapidly in early summer so that affected shoots are usually completely dead by late season, when differentiation of winter desiccation from other causes of die-back is extremely difficult.

4.6.3. Ramichloridium pini

Ramichloridium typically infects lodgepole pine shoots just behind the bud, causing shoot-tip die-back that progresses through the season to the death of entire one-year shoots. The fungus infects extending shoots early in the year but symptoms only become apparent in autumn. Infected shoots are most noticeable, and most easily identified, in late winter to spring (of the following year) when, typically, they have a cluster of yellow and red-brown needles at their apex (fig 45). Usually, the disease affects scattered shoots in the lower or mid-crowns of pole-stage trees, and stand edges may take on a multicoloured appearance as a result. In its earliest stage, discoloration progresses from the base to the tip of needles, whereas that caused by winter injury starts at the tip or affects the needles uniformly. Unfortunately, examination often takes place after discoloration is complete and so this may not be a useful distinction in practice. Buds on shoots infected by *Ramichloridium* may begin extension growth, but necrosis is nevertheless always visible in the bark – unlike shoots with foliage discoloration caused by winter injury. On heavily flowering provenances, the death of bark may begin around the flowers rather than at the shoot-tip. Severe

Fig 45. *Ramichloridium pini* on lodgepole pine, photographed in June. The two shoots on the right have typical symptoms resulting from infection of the shoot tip; the shoot on the left has not yet developed needle symptoms although the bud has failed to extend due to a lesion lower down the shoot.

infection by *Ramichloridium* may leave trees with only a tuft of live shoots at the very top. Repeated attacks of such severity can debilitate and even kill trees, but damage from this pathogen often looks worse than it is.

4.6.4. Brunchorstia pinea

Die-back caused by *B. pinea* is the most serious disease of pines dealt with in this section; severe outbreaks may destroy crops (fig 46). The fungus attacks one-year shoots of Corsican and Scots pine (figs 47 and 48) in much the same way as *Ramichloridium* attacks one-year shoots of lodgepole pine – infection occurs early in the growing season but remains latent until autumn or the following spring. The first symptom of *Brunchorstia* infection to be noticed is usually needle browning but, in contrast to winter injury, this is always the result of bark necrosis. Buds are also killed at an early stage of symptom development, another major distinction from winter injury. A third is that browning caused by *Brunchorstia* progresses from the base to the tip of needles, whereas the reverse is often the case with climatic damage. However, symptoms have to be caught in their earliest stages for this to be apparent and so it is often not a useful diagnostic feature in practice.

In Scots pine, infection usually starts in the lower crowns of late thicket or pole-stage plantations, where it may remain endemic, at a low level, on suppressed shoots. Under conditions favourable to the fungus, it builds up rapidly to cause severe die-back and mortality.

Fig 46. Scots pine plantation in which most of the trees are dead or dying as a result of *Brunchorstia* infection.

Fig 47. Corsican pine shoots killed by *Brunchorstia pinea*. [FC 21229]

The most serious attacks on Scots pine usually occur in 15–35-year-old crops on poor sites in the uplands. There is some evidence that provenance may influence susceptibility and it is noteworthy that no outbreak of this disease has been recorded in native pinewoods in Britain.

The disease is more predictable in Corsican pine since it is likely to cause serious damage on any except the best Corsican pine sites – namely those in the lowlands in the south and east of the country. The disease tends to attack crops after canopy closure even though they might have been vigorous and healthy in their early years.

Individuals that survive pole-stage can go on to make final crop trees, albeit more widely spaced than would be desirable. *Brunchorstia* infection also occurs in nursery stock and this may lead to die-back and failure on outplanting, even in areas where the disease is not normally a problem. On Corsican pine in their first year, death of buds, shoots and entire plants from *Brunchorstia* could be difficult to distinguish from frost injury (section 4.6.7) or planting failure (section 1.1).

Brunchorstia has only rarely been identified as a cause of damage on lodgepole pine in the UK and, for practical purposes, the disease can be disregarded on that species as it is currently used. Complacency could, however, be dangerous as the fungus is a serious problem on lodgepole pine in Scandinavia.

Fig 48. Corsican pine shoot killed by *Brunchorstia pinea*; the shoot tip has been sectioned to show bark necrosis extending below the level of needle symptoms. [FC 38163]

This disease cannot be diagnosed reliably from field symptoms on isolated shoots. In Scandinavia, the wood of shoots infected by *Brunchorstia* is described as acquiring a characteristic green colour but, in our experience, this is not such a common symptom under British conditions. Extensive shoot die-back in Scots or Corsican pine is unlikely to be anything other than *Brunchorstia* if abiotic damage (chemicals or climate) can be ruled out. This is sometimes difficult and so serious damage should be properly investigated – crops normally recover from climatic injury but they might not do so from *Brunchorstia* attack.

For academic reasons too tedious to go into, the fungus is known as *Brunchorstia* only in the UK; elsewhere it is referred to as *Scleroderris* or *Gremmeniella* and will be found under those names in some books. It may also be referred to as a canker disease, but this type of damage is rare under UK conditions.

4.6.5. Browning and Decline of Scots Pine

Three of the agents described separately in preceding sections, *Tomicus, Brunchorstia* and *Lophodermium,* can be jointly involved in outbreaks of needle browning and die-back that affect entire stands and deserve brief consideration under a single heading. Decline and die-back have mostly been observed in 15–25-year-old crops at higher elevation (more than 200–300 m) in northern Britain. The most serious factor has undoubtedly been *Brunchorstia* which, as already noted, is capable of killing trees in large numbers unaided. Even in lesser attacks, a high level of shoot loss may debilitate trees to the extent that they can be colonized and killed by *Tomicus. Lophodermium* probably contributes to crop deterioration if attacks are severe or repeated since severe defoliation may not only lead to die-back, but might also render trees liable to fatal attacks by *Tomicus.*

Tomicus may build up high populations in the dead and dying trees of declining stands and thereby become a significant cause of further crown reduction by shoot boring. In the later stages of decline, the insect may indeed be the only agent still in evidence.

Although significant mortality or total crop failure are possible outcomes (fig 49), browning and die-back in Scots pine may nonetheless represent an episode from which virtually complete recovery is possible. It is, therefore, worth investigating outbreaks with some care. If *Lophodermium* is the main agent responsible, the crop may not be at serious risk.

Lowland Scots pine plantations occasionally suffer forms of decline that are less dramatic than that instigated by *Brunchorstia,* though they can be equally destructive of productivity in the long term. There are no distinctive symptoms other than a general poor colour, thin crowns and loss of vigour. However, sporadic outbreaks of foliage browning may occur and, in most cases, a degree of die-back and mortality eventually develops in which a variety of pests and pathogens may be involved. It is likely that these attacks are more a reflection of, than an

Fig 49. Decline of Scots pine. Remnants of a once continuous stand in which the survivors (in front of Sitka spruce) are currently severely affected by *Lophodermium* needle-cast.

explanation for, the general poor condition of the crops in question. In many cases the underlying problem is probably an unsuitable site, or an unsuitable combination of site and provenance, and the most common limiting factor may well be poor drainage.

4.6.6. *Crumenulopsis sororia*

C. sororia can cause identical symptoms to those of *Brunchorstia* on Scots and Corsican pine – that is, death of one-year-old shoots – but it is only likely to affect scattered shoots in the lower crown. If dead or dying one- or two-year shoots have a resinous lesion at the junction of dead and live bark, the cause is likely to be *Crumenulopsis.* This kind of damage may occur in Scots, Corsican and lodgepole pines. As noted in section 4.5, the fungus can also cause resinous cankers on larger branches and main stems. These rarely seem to girdle as they do on small shoots and so the result is a long-lived canker that may induce copious resin flow and some deformation, but does not spread very far around the circumference. Indeed, despite the resinosis and deformation, there may in fact be little necrotic bark. The wood beneath lesions is often stained black, though it should be emphasized that while this may be characteristic, it is not specific to *C. sororia.*

4.6.7. Frost Injury

Though rare on Scots and lodgepole pine in Britain, frost injury is of some significance to Corsican pine. Small trees in their first two or three years after planting can be killed or checked by spring frost and on frosty sites repeated injury may seriously hamper establishment.

On trees up to 1 m or so in height, spring frost causes reddening of foliage and kills extending buds or shoots; soft shoots often collapse. Death of unflushed buds and spring die-back of shoots of the previous year is more likely to be due to *Brunchorstia* disease, though on small plants, especially those in their first year after planting, the two forms of injury could be difficult to distinguish – and either could be confused with planting failure. As noted in section 4.6.4, infection of planting stock by *Brunchorstia* may occur in nurseries but, otherwise, the disease is uncommon in the areas of lowland eastern Britain where Corsican pine is most favoured and where spring frost injury is most likely to occur.

5. DISEASES, DISORDERS AND PESTS OF LARCH

5.1. *Meria laricis*

This disease (fig 50), which is well known on European and hybrid larch in nurseries, occasionally causes striking defoliation in plantations. In both situations it could be mistaken for frost or chemical damage and, in plantations, it might also be confused with drought injury. The fungus attacks needles as early as May in some years, though symptoms

Fig 50. Needle-cast of hybrid larch caused by *Meria laricis*. [FC 39127]

are more commonly seen in early autumn. On infected needles, straw-coloured to pale brown bands may develop anywhere along the length of the needle but they are often concentrated at the tips, where they can look very like the effects of frost or herbicide damage. The occurrence of some needles with discoloration only of the base or middle, and a green tip, would suggest *Meria,* as would the presence of extension shoots with green needles left only at the tip. If all affected needles are completely brown, laboratory examination would be necessary for a confident diagnosis.

5.2. Larch Canker

"Larch canker" is the common name applied to the most serious fungal disease of larch: that caused by *Lachnellula willkommii.* This fungus has been placed in a variety of genera since its discovery and was known for many years as *Trichoscyphella willkommii.* The fungus infects young and old shoots causing resinous, perennial cankers that can girdle and kill side branches or leave major defects in main stems. Active cankers on still-live branches produce small fruit bodies, sometimes in a circle, on the flattened canker surface (fig 51). The fruit bodies are stalked discs up to about 3 mm across when open, with an orange-yellow centre and white, woolly margin.

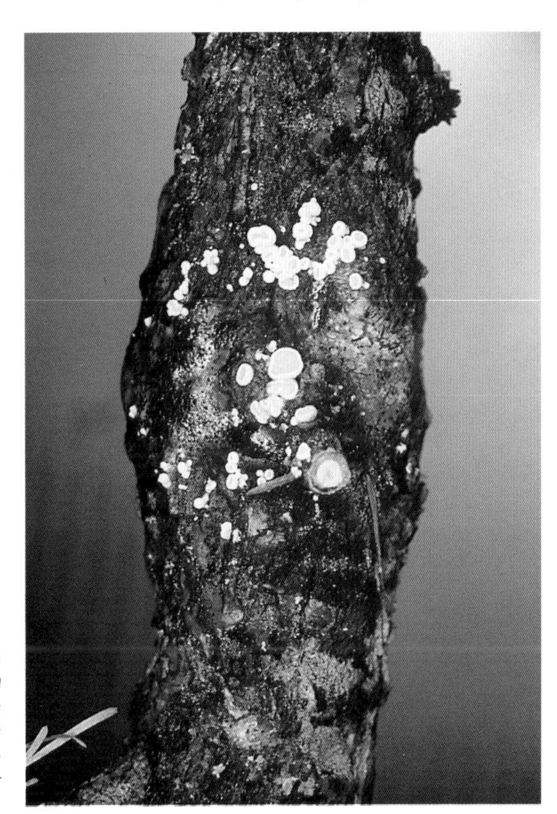

Fig 51. Larch canker with the causal fungus *Lachnellula willkommii* fruiting on the surface (a side branch at the centre of the canker has been cut off to aid photography).

Unfortunately, they curl up into undistinguished white lumps in dry weather (fig 52). Almost identical fruit bodies are produced by a related but non-pathogenic fungus that grows on dead twigs and branches and so it is important to examine active cankers on live shoots. On susceptible provenances of European larch, and on some origins of hybrid larch, this disease can cause such extensive cankering and die-back that stands may be rendered worthless. However, it is often difficult to separate the effects of "larch canker" from those of "larch die-back" described next (fig 53).

There is believed to be an association between frost damage and canker, such that bark injuries caused by the former may allow ingress of the fungus. Thus frosty sites may be particularly at risk. However, it should be emphasized that disease does not depend on frost injury; the most notoriously susceptible of all European larch provenances are alpine and extremely frost hardy.

Fig 52. Fruit bodies of *Lachnellula willkommii.* Note that these are partially (right) or completely (left) closed; in wet conditions they open as a flat disc. [Raymond Parks]

5.3. "Larch Die-back"

Trees susceptible to *L. willkommii* are usually also susceptible to a separate form of injury known confusingly as "larch die-back" – confusingly because "larch canker" itself also leads to die-back as described in the preceding section. "Larch die-back" (sometimes also referred to as "epidemic die-back") is induced by heavy attacks by the larch woolly aphid (*Adelges laricis*). Needles become smothered with aphid wool and sticky with honeydew. Eventually the whole crown takes on a dingy appearance as the foliage yellows, dark sooty moulds grow on the honeydew, and formerly vigorous shoots start to die back. Whether or not it is accompanied by canker disease, this disorder can cause serious damage to young larch (fig 53).

Fig 53. Hybrid larch severely affected by both larch canker and die-back. Photographed in spring following attack by larch woolly aphid in the previous year on trees already suffering from larch canker disease. The trees to the right have virtually all their shoots killed; the flushing tree to the left of centre is largely unaffected.

5.4. *Ips cembrae*

All three larches can be killed following attacks by the bark-boring beetle, *Ips cembrae*. This insect carries a pathogenic fungus, *Ceratocystis laricicola*, which is introduced into the tree during breeding attacks on the main stem bark. Development of the fungus causes lesions which, after multiple beetle attacks, overlap each other and girdle the stem, thereby killing the top.

Trees under stress are preferentially attacked and, as the insects breed in moribund material, trees near fresh larch logs from recent fellings may also be at risk from locally high populations. A combination of stress and nearby breeding material is often a feature of remnant crops left exposed on the fringes of clear-felled or windthrown stands. Larch in this situation is vulnerable to attack, particularly after a prolonged spell of dry weather.

Trees killed by this association between beetle and fungus are characterized by the presence of insect galleries under the bark – there are no other gallery-forming insects likely to attack newly killed larch. However, the activity of the insect may not be evident in the

Fig 54. Top of European larch killed by *Ips cembrae* and *Ceratocystis laricicola*.

lower stem of large trees since it often attacks first near the base of the live crown or even higher. Sometimes resin dribbles down the lower stem from the beetle attacks higher up. The presence of a few trees with dead tops but live lower crowns (fig 54) is characteristic of this problem in older stands.

As noted in section 1.4.4, larches are particularly vulnerable to drought, which may lead to top die-back without the intervention of *Ips cembrae*. It is also relevant to note here that the insect is not yet known to occur in southern Britain.

5.5. Minor Causes of Shoot Die-back in Larch

There are three agents that are capable of killing isolated shoots, up to about three years old, in larch but that do not go on to cause wholesale die-back or decline. One is the larch shoot moth, *Argyresthia laevigatella*, whose involvement is made obvious by the presence of a bore-hole or tunnel at the junction of dead and live bark. Attacked shoots usually die just before or shortly after flushing, and the junction between dead and live parts itself becomes obvious by the abrupt increase in diameter of the still-growing part.

Fig 55. Larch shoots killed by *Phacidium coniferarum*.

Death of shoots early in the season with the same symptoms as damage by the shoot moth but without the tunnel is likely to be the result of fungal infection, either by a *Cytospora* species or by *Phacidium* (*Potebniamyces*) *coniferarum*. Both have been associated with the occurrence of flagging shoots (fig 55) that may be widespread and extremely noticeable, though probably not seriously damaging, on all three larches in spring and early summer. *P. coniferarum* can attack a wide range of conifers causing minor canker and die-back diseases. It is probably best known as the cause of die-back in young Douglas fir, especially after initial injury by frost – see section 1.4.1.

6. DISEASES, DISORDERS AND PESTS OF FOREST BROADLEAVES

6.1. General Remarks on Diseases, Disorders and Declines of Broadleaves

For a comprehensive coverage of the pests and diseases of broadleaves, readers are referred to "Diseases of Forest and Ornamental Trees" by D H Phillips and D A Burdekin (Macmillan; 2nd ed; 1992), "Diagnosis of Ill-health in Trees" by R G Strouts and T G Winter (HMSO; 1994) and "Forest Insects" by D Bevan (Forestry Commission Handbook 1; HMSO; 1987). In the key we have covered only a selection of the most common problems on species likely to be used in commercial plantations. Moreover, we have concentrated on problems associated with fairly clear above-ground symptoms. Root diseases of forest broadleaves are difficult to diagnose and the one most likely to be encountered, *Armillaria*, may well not be primary, but an indication of a more fundamental disorder. With the exceptions of *Prunus* species and willows, most native broadleaves are resistant to *Armillaria* attack.

Ground disturbance and other changes in the rooting environment can impose sufficient stress on large, old broadleaves to precipitate a protracted decline. Root-infecting fungi are often involved in such declines since the initial stress or injury can allow them to establish a potentially fatal foothold in the root system. *Armillaria* is the most common of these opportunists, but in beech, which is particularly susceptible to ground disturbance and drought, two other root-disease fungi, *Ganoderma* species and *Meripilus giganteus*, are also often involved. The former produces thick, woody, perennial brackets attached to the stem base, while the latter produces large (up to 30 cm across) fan-shaped fruit bodies in autumn on the ground close to the base (fig 56). Though substantial, *Meripilus* fruit bodies are rather fleshy and soon discolour and deteriorate after frosts. Both fungi can act as primary root pathogens and both can render trees unsafe, though *Meripilus* is particularly dangerous in this respect. Beech, like sycamore and birch, may also develop above-ground infections as well as root infections in response to drought stress.

Fig 56. Beech with a clump of *Meripilus giganteus* fruit bodies at the stem base (left); a *Ganoderma* bracket is also present on the right.

Ash and oak are subject to long-term decline and die-back for which living agents are not thought to be responsible, although they may play a supporting role. The causes of "oak die-back" (fig 57) and "ash die-back" are not known, though both are believed to be stress-related disorders. Each might, in fact, be instigated quite independently by a number of factors, of which drought has probably been the most widely discussed, especially in relation to the serious outbreak of "ash die-back" that occurred in East Anglia in 1992. It is worth emphasizing that decline and die-back disorders of large trees have really quite general symptoms – crown thinning, yellowing, twig and branch die-back – which culminate either in death or in a "stag-headed" condition from which the tree may eventually recover by epicormic growth (fig 58). The same symptoms can be induced by other problems – root disease and waterlogging, for example. Only when die-backs are widespread and investigation has failed to find plausible local causes for individual cases is it appropriate to consider the possibility of a general syndrome precipitated by large-scale environmental influences such as climatic cycles or pollution.

Fig 57. Early symptoms of "oak die-back".

Fig 58. "Oak die-back"; a stag-headed tree with epicormic growth.

6.2. Insects and Mites

Most attacks on the foliage of broadleaves by insects lead to physical damage whose general cause is quite obvious even if the particular insect responsible is not. However, there are some exceptions worthy of mention here.

On beech, infestations by the leaf miner *Rhynchaenus fagi* cause brown blotches at the tip or around the edges of leaves. When held up to the light, these patches are transparent, quite clearly mined and usually contain frass. However, from a distance the symptoms look very like climatic or chemical injury, especially as these insects can be so numerous as to affect nearly all the crown.

A variety of minor leaf deformations are induced in broadleaves by eriophyid mites. These cause small spiny or felty outgrowths, or pimples, on leaves of beech, elm, sycamore and lime among the more common forest broadleaves. Occasionally these outgrowths are sufficiently numerous to cover the whole leaf surface.

6.3. Rusts

These are the most easily recognized foliage diseases. Rusts on broadleaves usually produce yellow or orange pustules (fig 59) in

Fig 59. Rust (*Melampsora* sp.) on poplar. [FC 40787]

early or mid-summer and brown pustules in late summer and autumn. Pustules usually form on the underside of the leaf blade (though visible as spots on the upper surface), but in some species they appear on both sides. The earlier pustules produce spores that re-infect the summer host while the later ones produce spores that infect an alternate or over-wintering host (which may be another tree species). This alternation of hosts is not obligatory for all rusts as some – for example, birch rust – can overwinter successfully on fallen leaves or on twigs and buds of the main summer host. Rust diseases are strongly influenced by climate so that the amount of infection varies from season to season. Severe attacks may cause infected leaves to die and fall prematurely. Rusts are highly specialized parasites and cannot survive outside the living host, features that make them extremely difficult to control, even through the breeding of resistant cultivars. The most important rusts of forest broadleaves are as follows.

Poplar rusts. These are caused by several species in the genus *Melampsora* (fig 59). Serious infection in commercial, supposedly rust-resistant, clones would merit specialist advice. Rusts do not deform the leaf blade; yellow blisters on poplar leaves are likely to be caused by another fungus, *Taphrina populina*.

Birch rust *(Melampsoridium betulinum)*. Severe attacks may cause such heavy spotting that without close examination it is not clear that a rust is the culprit. Heavy attacks give the infected trees an overall dingy yellow appearance and are followed by premature defoliation from the lower crown upwards. There is great genetic variation in susceptibility and even in bad years some individual trees may be relatively little affected next to heavily rusted neighbours. Where birch is a dominant component of the treescape (as in parts of the Highlands), bad rust years can give an impression of early autumn.

Willow rust. Like poplar rust this is caused by several species in the genus *Melampsora*. Infection may reduce the productivity of short-rotation coppice crops and so this disease is of some commercial importance.

6.4. *Apiognomonia errabunda*

Foliage browning on beech and oak in the form of large blotches at the leaf margin or between the veins can be caused by *Apiognomonia errabunda* (fig 60). In seasons favourable to the fungus, it may cause severe defoliation in oak, and may lead on to twig die-back. In both species, microscopic examination is necessary to confirm the presence of the disease. On oak, attack by the aphid known as oak leaf phylloxera (*Phylloxera glabra*) can lead to severe blotching which superficially resembles that caused by *Apiognomonia*.

Fig 60. Oak leaves with spots and blotches caused by *Apiognomonia errabunda*. [FC 40892]

6.5. Leaf Spotting and Browning of Sycamore

Sycamore is subject to a number of foliage diseases. The commonest is the well known "tar-spot", the name of which provides a graphic description of the symptoms caused in late summer and autumn by the fungus *Rhytisma acerinum*. Initially, the spots are yellow, and a yellow halo is usually retained round the final raised, black "tar-spots". In some years infection can be extremely heavy and lead to premature defoliation. *Rhytisma* may be accompanied by two other fungi, *Cristulariella depraedans* and *Phleospora aceris*, which cause whitish or greyish brown, unthickened leaf spots. The last two fungi require laboratory examination to diagnose. A fourth fungus, *Ophiognomonia pseudoplatani*, produces large blotches on leaves, often at the base of the leaf blade where it joins the stalk. However, some care is necessary with this symptom as wind may cause similar damage. Sycamore leaves present a large sail area and strong winds in the growing season can cause physical damage and collapse of tissue at the junction of leaf and stalk. This in turn leads to discoloration spreading out from that point as the leaf desiccates.

Sycamore leaves can also suffer marginal browning from wind damage, though this symptom also occurs quite often in situations where wind exposure or chemical injury (a possible cause of marginal browning in broadleaves generally) do not seem likely explanations. Waterlogging may lead to marginal browning in sycamores, though usually there are other signs of foliage distress as well. Attack by the sycamore aphid has been proposed as the reason for otherwise unexplained cases of marginal browning, but there is no direct evidence for this. Frequently, the majority of leaves on some individuals are damaged while neighbouring trees are unaffected

6.6. Diseases of *Prunus*

6.6.1. Foliage Diseases

Cherries and plums are subject to several diseases, some of which are extremely serious while others are dramatic but probably have little long-term significance. The foliage diseases caused by *Apiognomonia erythrostoma* and *Blumeriella jaapii* fall into the second category. They can only be diagnosed with certainty by identification of the causal fungi on the leaf surface, though their symptoms do differ. *A. erythrostoma* (cherry leaf scorch) is common on gean in some years. Late in the season it produces large reddish brown blotches with yellow halos on infected leaves. The most characteristic feature of the disease is that the shrivelled leaves eventually hang down and remain attached to the shoots throughout winter. However, *Sclerotinia*, a die-back disease (see next section), also leaves withered foliage attached and so some care is needed to establish whether the shoots bearing the leaves are alive or dead. *Blumeriella jaapii* causes small spots which are purple on the upper side and brown on the underside. The centres may eventually fall out leaving a "shot-hole". The leaf-infecting phase of bacterial canker also causes "shot-holing" and so there is a possibility of confusion with this serious disease (see next section).

6.6.2. Die-back Diseases

There are three diseases that can cause serious die-back in *Prunus* – "silverleaf", bacterial canker and *Sclerotinia* – all of which can be difficult to separate from each other on field characters.

"Silverleaf" is so-called because the foliage of affected parts of the crown assumes a silvery or leaden sheen. This symptom is often followed by protracted die-back and decline. Fruit bodies of the causal fungus – known as both *Stereum purpureum* and *Chondrostereum purpureum* – may form on the bark of dead limbs at a late stage of the disease. The fruit bodies are small (2 or 3 cm wide) bracket-like structures with a purple underside when actively growing. Silverleaf disease is most likely to be encountered on plum, but also affects gean. Since the causal fungus can infect trees through pruning wounds, pruning should be carried out in summer (June–August) when the risk of infection is minimal.

Bacterial canker, which is common on gean, also affects both cherries and plums. Although the bark lesions characteristic of this disease develop during the dormant season, their effects are usually noticed when individual shoots, or quite large branches, wilt and die back early in the growing season. Lesions may be recognizable as

Fig 61. Gum oozing from a lesion caused by bacterial canker disease on the stem of a young gean; the canker is visible as a slightly depressed and ridged patch above and below the junction of the side branch.

depressed patches in the normally shiny bark of cherries, but they are often more readily identifiable by the gum that oozes from them, sometimes in copious quantities (fig 61). During summer, the bacteria infect leaves to cause small spots which later drop out to leave "shot-holes". Infection of branches and stems takes place through leaf scars or bark wounds in autumn or winter. As serious damage can follow infection of pruning wounds, pruning should only be carried out in summer. The bacterium (or possibly a group of related bacteria) responsible for this disease is variously referred to as *Pseudomonas mors-prunorum*, *P. syringae* and *P. syringae* pathovar *mors-prunorum*

Sclerotinia (Monilia) laxa, a fungal pathogen, infects leaves and flowers in spring, causing them to darken and shrivel. It then passes into the shoot where a dark, sunken lesion develops. Spur shoots on cherries may wilt rapidly as a result of infection, and the disease may then spread into larger branches, when it is difficult to distinguish from bacterial canker. Die-back caused by *Sclerotinia* can affect the whole crown of ornamental cherries, but infection is not normally serious on species likely to be found in woods and plantations.

6.7. Bacterial Canker and Other Poplar Diseases

In the past, poplar canker caused by the bacterium *Xanthomonas populi* was probably one of the most serious known diseases of plantation broadleaves, but modern commercial poplar clones are screened against it. Cankers usually appear in spring or early summer as sunken or cracked areas of bark on shoots of the previous year. The young cankers may exude whitish bacterial slime in wet weather. Older cankers usually have a sunken centre (sometimes with exposed wood) and a broken, irregularly swollen margin. They may also develop into rough erumpent growths. On susceptible trees, the disease can lead to substantial branch die-back, while main stem cankers may cause serious timber defects.

Aspen and some other poplars are subject to scab diseases caused by *Venturia* species, relatives of the more common willow scab (section 6.11). Large black leaf blotches develop in spring and early summer on infected leaves, which eventually become completely blackened and shrivelled. The pathogen quickly spreads into young shoots causing them to wilt, blacken and die. The blackening of infected shoots and leaves can be quite striking in severe cases, though serious die-back is rare in the UK other than on aspen.

On poplars other than aspen, the most common defoliating diseases are those caused by *Melampsora* species (rust – see section 6.3) and *Marssonina* species. Infection by the latter starts as small brown spots, but may quickly discolour the whole leaf. Infection and defoliation usually start at the base of the crown and work upwards so that, in severe attacks, only a tuft of foliage remains at the top of the crown by late summer.

Venturia and *Marssonina* diseases require laboratory examination for a confident diagnosis. Both can cause such severe symptoms that they could be mistaken for serious climatic or chemical damage.

6.8. Beech Bark Disease

Beech bark disease is caused by the interaction between a pest, the felted beech coccus (*Cryptococcus fagisuga*), and a pathogen, the fungus *Nectria coccinea*. The coccus is a bark-feeding insect whose activities injure cambial cells and permit invasion by the fungus. The latter extends the damage and may go on to cause large, and often fatal, stem lesions. The presence of the coccus can be recognized by flecks of white wool (figs 62 and 63) produced as a protection against predators. In heavy infestations, trees look whitewashed, but less obvious attacks may nevertheless be sufficient to instigate beech bark disease. Conversely, heavy attacks do not inevitably lead to disease, though the stems of trees that recover may have a dimpled appearance resulting from bark injury by the insect (fig 62). On diseased stems young lesions are usually recognizable by the exudation of dark fluid ("tarry spot" – fig 63); older lesions are revealed by detached bark. By the time that bark symptoms such as dimpling or tarry spots appear, the coccus population is often quite

Fig 62. Beech stem heavily dimpled from attack by felted beech coccus, which is still evident as patches of white "wool" [FC 37561]

low and inconspicuous. Trees with areas of dead bark are often attacked secondarily by ambrosia beetles and decay-causing fungi, one of which, *Bjerkandera adusta*, appears to be particularly suited to establishing itself in the galleries made by the beetles into the stem wood. Invasion by *B. adusta*, and similar fungi, can lead to early breakage ("beech snap") of diseased stems, typically at about three to five metres above the ground.

Fig 63. Beech bark disease; "tarry spots" caused by *Nectria* infection. White woolly deposits produced by felted beech coccus are present on the right of the stem. [FC 37560]

Outbreaks of beech bark disease often start in even-aged plantations between 15 and 25 years old, and then usually reach a peak around 20–30 years of age. During this period losses may be significant, but are unlikely to reach a level that threatens a final crop. Indeed, they may not greatly exceed the normal volume of early thinnings. After the peak phase the disease tends to decline, though limited mortality can occur throughout the life of a stand.

There is no clear evidence to support silvicultural intervention as a means of disease management. It may be worthwhile to salvage heavily infested trees, whose annual increment is likely to be extremely low, but moderately or lightly affected individuals may well recover and should not be selectively removed. Studies in old, formerly diseased, stands suggest that, despite some gaps, the overall stocking density and volume are comparable to expected yields. However, quality might suffer in gappy crops, or in crops where disease losses have pre-empted the selection of trees for removal in early thinnings.

It is worth commenting that the pathogen involved in beech bark disease, *N. coccinea*, is adapted to attack stressed bark, and that causes of stress other than infestation by the coccus can allow it to establish a foothold. One such cause is drought, to which, as noted in section 1.4.4, beech is susceptible.

6.9. Dutch Elm Disease

Although the greatest impact by Dutch elm disease in Britain has been the widely publicized devastation of roadside, hedgerow and urban populations of elm, it is nevertheless also a serious disease of woodland elm. While great damage has been done by this disease in central and southern England, there are still substantial populations of elm in northern and western Britain, and even significant surviving pockets in areas like East Anglia. Managers of lowland broadleaves can, therefore, expect to encounter the disease, even in the areas where it is currently rare or unknown (most of the Highlands and Islands, Argyll and Grampian north of the Don valley).

Elm disease is caused by one of two fungi, *Ophiostoma ulmi* and *O. novo-ulmi*, which belong to a group of pathogens known as "vascular wilts". As the name suggests, they cause disease by infecting, and eventually blocking, the water conduction (vascular) system of plants. The elm disease pathogens are spread by elm bark beetles (*Scolytus* species) which breed in the bark of dead or dying elms. Young beetles emerge from these trees in spring and fly to the crowns of healthy trees where they complete their maturation feeding by excavating grooves and tunnels in the living bark of young shoots. If breeding takes place in trees affected by elm disease, the emergent beetles are likely to be contaminated by *Ophiostoma* spores, which may then be transferred to the feeding wounds in healthy trees. Once established in the young wood, the pathogen spreads through the conduction system into the main limbs and then usually into the main stem. Leaves on affected branches wilt, turn yellow, then brown, and fall. Because the disease moves progressively through infected trees, they often exhibit a mixture of normal foliage, yellow or brown foliage, and defoliated branches (fig 64). Symptoms typically appear from late June onwards and the trees are usually dead by the following spring. The more common of the two pathogens, *O. novo-ulmi,* is the more aggressive and can spread through a mature tree in weeks. The presence of either fungus in live twigs is marked

Fig 64. The effects of Dutch elm disease in wych elm at the edge of a small wood. The combination of defoliated branches and large patches of yellowing and browning foliage in the tree behind the cow is typical of the disease in woodland wych elm; the dead tree to the left is an earlier victim.

by brown or purple longitudinal streaks in the outermost, infected wood. They can usually be seen by peeling the bark from symptomatic, live twigs and are diagnostic for elm disease.

Discoloration and die-back of large parts of the crown in woodland elm, when other species remain healthy, is unlikely to be anything other than elm disease (fig 64). The same presumption cannot be made in the case of stag-headed trees and trees with only a few small, dead shoots in the crown. In these, and in suspect trees outside the main area of distribution of the disease, the confirmatory streaks in the outer wood of peeled twigs should be sought.

Control of the disease by sanitation felling is still practised in some areas. In these, the local authority is empowered to inspect premises, including woodland, and compel owners to deal with diseased trees at their own expense by a due date. The action usually required is to fell, debark and then burn the bark and branchwood, a procedure aimed at denying breeding material to the elm bark beetle. Sanitation felling is not recommended outside statutory control areas and, since 1993, there has been no statutory restriction on the movement of elm wood – owners and merchants should, however, be encouraged not to move elm from infected areas into or across control areas.

There is little immediate prospect of elm disease declining. Its apparent absence from some old outbreak areas merely reflects the absence of elms large enough to support a beetle population. As soon as regrowth from suckers, or seedlings in the case of wych elm, reaches any size, the beetles and the disease reappear. As long as this situation prevails, elm is not recommended for planting anywhere in Britain.

6.10. Sooty Bark Disease of Sycamore

Sooty bark disease of sycamore is caused by the invasion of wood and bark of susceptible trees by the fungus *Cryptostroma corticale*. The wood of affected branches develops a yellowish green stain, but the outward symptoms are wilting (if the trees are in leaf), die-back and, most characteristically, the production of *Cryptostroma* fruit bodies. These form between the wood and bark as a layer of fungal tissue that causes the bark to blister and flake off to expose the sooty black sporing surface. The disease is usually fatal and in dead or dying trees the fungus may produce large areas of "sooty bark" on the main stem (fig 65). It is likely that *Cryptostroma* can inhabit healthy sycamores without causing symptoms but is stimulated into pathogenic activity by severe stress in the host. This form of behaviour, known as latent pathogenicity, is exhibited by a variety of tree pathogens (see also section 6.12). Outbreaks of sooty bark disease are probably triggered by a combination of high summer temperatures and drought, though symptoms usually develop in the following year. Despite the dramatic nature of this disease, it is not easy to diagnose and is frequently blamed

Fig 65. Sooty bark disease of sycamore; the black patches are spore-producing surfaces of *Cryptostroma* fruit bodies that have forced off flakes of bark. [FC 29877]

for other problems. There are a number of fungi with black spores or black fruit bodies that will act as secondary invaders of sycamores killed by other agents – even *Cryptostroma* will behave like this. Because of the importance of high temperatures and drought stress in its development, sooty bark disease is most prevalent in the southern half of England; reports of it from elsewhere should be treated with caution.

6.11. Shoot Diseases of Willow

The two shoot diseases of willow known as scab (caused by *Venturia saliciperda*) and black canker (caused by *Glomerella miyabeana*, also known as *G. cingulata*) are indistinguishable on field symptoms and for practical purposes may be treated as a single disease. Both cause large black leaf blotches (progressing to complete withering), black lesions on shoots, and die-back (fig 66). Severely infected young trees are quite distinctive since they look as if they have been sprayed with a contact herbicide. However, it is possible to confuse lesser symptoms with those of another fungal pathogen, *Marssonina salicicola*, which also causes leaf spots and lesions on young, green shoots. This is most common on weeping varieties, and other varieties with a blackened, blasted look are most likely to be affected by scab/black canker.

Fig 66. Crack willow showing extensive crown damage caused by scab disease in the previous year; recovery shoots are evident on some affected branches.

From a distance, shrubby willows may also look scorched when they have been defoliated by leaf beetles. At close range the effects of the latter are quite unmistakeable as there is no damage to shoots while the leaves, although not shrivelled, are reduced to papery skeletons.

The important commercial variety, cricket bat willow, and some ornamental willows are subject to a bacterial disease known as "watermark" from the stain it causes in timber. The disease often causes die-back, though not invariably. Tree willows with reddening and shrivelling of leaves on one or more branches could be infected by watermark disease, but specialist advice would be necessary to confirm this.

6.12. Bracket Fungi and Death of Birch

Death of older birch trees is frequently associated with the production of large fruit bodies on the dead or dying stem. Such trees may comprise a high proportion of declining birch stands where regeneration is inhibited by browsing pressure from deer, rabbits or domestic stock. The fungus most regularly involved is *Piptoporus betulinus* – the common birch polypore – which produces distinctive, leathery "brackets" up to about 25 cm across and more or less semicircular when seen from above or below. The colour of the top varies from chestnut to greyish brown, while the lower, poroid surface is white or off-white. In Scotland, *Fomes fomentarius* is also common on birch, producing grey, very hard, hoof-shaped brackets – though for these the American term "conk" seems more appropriate than "bracket". A third fungus, *Inonotus obliquus,* can also be found on the stems of declining birch in Scotland, but its fruit body, an irregular, erumpent black crust which is barely recognizable as a fungus, is much less conspicuous than the other two.

These fungi appear not to be aggressive pathogens and their role in the decline or death of birch trees is problematical. Although they may only have a limited ability to attack healthy trees, it seems likely that they are effective invaders of stressed trees. In the case of *P. betulinus,* there is evidence that it functions as a latent or endogenous pathogen. This term is used for organisms that can become established in healthy tissues but remain almost inactive and do not elicit disease – a type of symptomless infection. However, should the host be placed under stress, the latent pathogen is well placed to break out and achieve rapid and widespread colonization of the weakened tissues. In the case of stem pathogens, this may trigger bark lesions and die-back – sooty bark disease of sycamore, described in section 6.10, is an example. Whether invasion occurs from outside or inside, it is probable that the major predisposing stresses in birch are senility and periodic adversities of site or climate such as drought, waterlogging or storm damage.

6.13. Ink Disease of *Castanea* and Other *Phytophthora* Diseases

Ink disease is probably the most common cause of death in *Castanea* where it is grown commercially in the UK. It is a root disease caused by species of *Phytophthora* (*P. cinnamomi* and *P. cambivora*) and so the crown symptoms are like those induced by other fatal root problems: thinning, yellowing, production of small leaves, late flushing, premature autumn leaf-fall, decline and die-back. These symptoms are associated with the death of roots and often, as infection spreads up from the root system, with the appearance above ground of inverted-V-shaped lesions at the stem base (fig 67). Such lesions are probably the single most useful diagnostic feature of the disease in large trees. In young coppice, the disease can cause rapid flagging of sprouts but it is characteristic that only a proportion on any one stool is killed. Because of the more indirect connection between root and stem in young sprouts, they are less likely to develop basal stem lesions than older trees or trees originating as seedlings. The name "ink disease" derives from

Fig 67. *Castanea sativa* with a basal lesion caused by *Phytophthora cambivora*. Bark has been pared away around the margin of the lesion. [FC 36897]

the blue–black colour taken on by the dead stem and root tissues, and from the inky fluid that sometimes weeps from them. However, this is not a completely reliable symptom as quite healthy wood and bark of *Castanea* may produce this stain from contact with minerals in the soil.

Root diseases caused by *Phytophthora* species are well known and extremely damaging on ornamental trees but, other than in *Castanea,* they have made virtually no impact under plantation conditions in the UK to date. However, they could do so, especially in the south of the country. Of most immediate concern is a recently discovered *Phytophthora* disease of common alder which, in addition to attacking riparian alder, has caused significant mortality in a number of flood-plain plantations. *Alnus incana* and *A. cordata,* as well as common alder, have been affected.

The circumstances in which a *Phytophthora* disease might be suspected are the death of several young trees, apparently from root disease, in wet ground conditions, but where straightforward waterlogging seems unlikely. These diseases are extraordinarily difficult to diagnose and require specialist investigation.

7. CAUSES OF PHYSICAL DAMAGE

7.1. Damage by Humans, Other Mammals and Insects

Most forest managers are all too familiar with the appearance and consequences of physical damage caused by insects or mammals. The attribution of damage to a particular animal is not always straightforward and the advice of an experienced ranger should be sought by anyone faced with an unfamiliar type of damage or by an unusually severe outbreak of an, apparently, familiar type. Fraying by deer is nearly always obvious from the buckled and hanging strips of bark on young trees – though it may require close examination to make sure of these features. When investigating physical injuries on trees it is important to remember that they are unlikely to lead to crown symptoms unless they completely, or very nearly, girdle the stem.

Where bark has been removed from trees in their first few years, the most likely culprits are *Hylobius* weevils or voles. On sites where both are a risk, it might not be easy to distinguish their damage without close examination and some experience. Voles are capable of girdling the bases of trees several centimetres in diameter, when their activities could be confused with those of larger mammals. Rabbits, hares, deer, squirrels and domestic stock are capable of girdling the bases of large trees, particularly thin-barked species.

Although bark stripping above about 2 m is most likely to be due to squirrels, in some areas edible dormice or bank voles, both of which climb, could be involved. Some weevils, including *Hylobius*, can also remove small patches of bark from higher branches. In sycamore and beech crowns, squirrel damage may be so severe that the obvious can be overlooked in favour of more sinister explanations. Squirrels typically seek out the inner bark and discard the strips of outer bark that they have to remove first. Unless the pieces are fresh, they may give the appearance of dead bark that has dropped from branches attacked by a pernicious disease. Squirrels engage in other forms of feeding that give rise to equally confusing signs. In early summer they often chew partly through young branches of horse chestnut (and occasionally ash) to extract the pith. The branches wilt and may break at the point of attack so that they hang down with brown leaves to give the appearance of a shoot disease. Outwardly similar damage, but distinguishable by the presence of a bored tunnel along the whole affected length of the shoot, may result from attack by the leopard moth on horse chestnut.

Girdling injuries by winch anchor cables are an occasional cause of death in large trees in commercial stands.

Damage to the wood surface as well as the bark usually indicates a human- or machine-made wound. However, badgers can score tree bases deeply and sika deer occasionally score stems higher up with their tines.

Some insects cause die-back or death of trees by tunnelling into bark or wood. The best known of these are the bark beetles *Tomicus piniperda*, *Ips cembrae* and *Dendroctonus micans*, and the weevil *Pissodes castaneus*, which can tunnel into both wood and bark of young pines. Because of the threat presented by several highly damaging non-indigenous species, we recommend that specialist advice be sought in any cases of apparent attack by bark-boring insects on conifers.

7.2. Damage by Non-living Agents

The only common type of physical damage by a non-living agent is stem crack caused by drought in conifers. These usually spiral round the stem and may heal rapidly or continue to open and close for some time. Drought cracks occur suddenly with a noise like a shot and can penetrate quite deeply into the wood. They usually involve only a simple split, without the bark death and scarring that result from lightning damage (section 1.4.5). On appearance alone, drought cracks could be confused with frost cracks, but the latter rarely occur in Britain.

8. STAIN AND DECAY

8.1. General

The main preoccupation with stain and decay in plantation trees is potential loss of produce. However, in large trees, decay also has safety implications which forest managers have increasingly to take into account as the use of forests for recreation increases. Those interested in tree safety are recommended to read "Principles of Tree Hazard Assessment and Management" (Lonsdale in press). It should be emphasized, in consideration of both safety and volume loss, that advanced stem- or root-rot is not necessarily associated with crown deterioration or any other outward sign of damage.

There are two major sources of stain and decay in standing trees: invasion of wounds, including pruning wounds, and attack by root pathogens, several of which can subsequently spread into the stem as butt-rots. It is relatively easy to establish whether a decay column is associated with a stem wound and, although the degree of examination necessary to do this may be time consuming, it is worthwhile where there is any degree of loss involved in commercial stands. Wound-rot, no matter how severe, would not prejudice the next rotation, whereas the presence of a butt-rotting fungus like *Heterobasidion annosum* could do so.

8.2. Wound-rot

The outer layers of bark on tree stems contain materials that make them extremely intractable to micro-organisms so that, together with the living cells of the inner bark, they make up an effective barrier to pathogens. In healthy trees, breaches in this barrier usually elicit vigorous responses involving healing or compensatory growth and the release of protective substances into the exposed wood. From the standpoint of commercial timber production, these reactions are themselves responsible for significant defects such as stem deformation, resin-soaking and discoloration. Despite defensive responses, a proportion of wounds will inevitably allow the ingress of stain- and decay-causing micro-organisms (fig 68). In some species – Norway spruce and beech, for example – this proportion can be high. As a rule of thumb, the larger the wound, the higher the risk of serious degradation.

In addition to wounds made by humans, machines or animals, drought cracks can allow the entry of stain and decay if they remain open. In crops where obvious cracks occur, it is worth making a careful inspection at the felling of apparently unaffected trees since there may well be individuals with small cracks that healed so rapidly that they became barely noticeable. Despite their small size, these could be numerous enough to represent a significant defect at conversion.

Fig 68. Wound rot in Sitka spruce. The original wound surface, parts of which can be seen in the lower half of the section, has been obscured by healing growth pushing into the rotted area.

A large number of fungi are able to invade wounds to cause stain and decay, and for most practical purposes in forestry it is unimportant to distinguish one species from another – though in the field of arboriculture, some current research is aimed at elucidating the special risks associated with particular species. If the fungus responsible for decay produces a fruit body on the wound surface, identification could be attempted from a standard mycological text or from the descriptions of the most common species given in: "Diagnosis of Ill-health in Trees" by R G Strouts and T G Winter (HMSO; 1994); "Diseases of Forest and Ornamental Trees" by D H Phillips and D A Burdekin (Macmillan; 2nd ed; 1992); "Principles of Tree Hazard Assessment and Management" by D Lonsdale (The Stationery Office; in press).

One species, *Stereum sanguinolentum*, merits special note as probably the most common and damaging cause of wound-rot in commercially grown conifers in Europe. It attacks a wide range of species and can colonize stems rapidly – rates of advance of decay up to 20 or 30 cm/year have been associated with it. It produces light-coloured and insignificant bracket-like fruit bodies which, if they are in good condition, turn red when cut or bruised. The fungus is ubiquitous and colonizes woody debris as well as standing conifers.

There is no equivalent of *S. sanguinolentum* on broadleaves. A wide range of fungi, often associated with particular host species, can cause serious decay.

8.3. Butt-rots

By far the most important agent of butt-rot in plantations is *Heterobasidion annosum* (Fomes). It affects most commercially grown conifers but, although it occasionally attacks broadleaves, it is a negligible problem in them. Unfortunately, in view of its potential for causing volume loss (fully covered in section 2.3), Fomes butt-rot can be difficult to diagnose in the field, though its presence is usually fairly easy to establish in the laboratory. Infection is usually associated with staining which, in newly infected trees, may be a pink or brown patch on one side of the butt roughly mid-way between the pith and the cambium. In trees with long-standing infection, the whole heartwood may be affected, but the discoloration may actually be less obvious. The colour and physical appearance of decay varies between species, but the inner part of the heartwood is usually light brown, with a pinky purple discoloration in the outer heartwood. Advanced decay (fig 69) is usually dryish and light brown with white flecks, though these are not always easy to see. In some species, including Sitka spruce, the stain associated with *H. annosum* can be difficult to distinguish from the natural coloration of the heartwood. It is worth commenting that, whereas the strong heartwood coloration of species such as Douglas fir and the larches is well known, it is not widely appreciated that Sitka spruce heartwood may show similar features. It varies from white, through pink, to almost

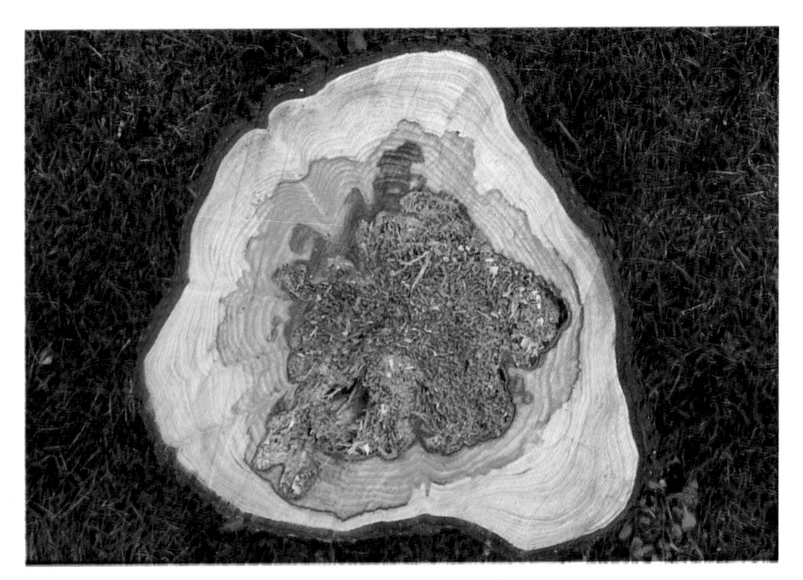

Fig 69. *Heterobasidion annosum butt-rot in Tsuga heterophylla.*

reddish purple (fig 70) and, like the stain caused by *H. annosum*, it is usually most strongly developed in the outer part of the heartwood and at the base of the tree. Strong coloration is more common in Sitka spruce on dry mineral soils than on wet peats, but it can be found on any site type.

Fig 70. Natural heartwood coloration in Sitka spruce.

As a consequence of its life cycle, described in section 2.3, *H. annosum* is extremely unlikely to occur in first rotation, unthinned crops. First rotation crops that have received stump protection since the time of first thinning are also likely to be substantially free of the disease. However, in first rotation crops thinned without stump protection, and in all subsequent crops, any butt-rot that is not a "brown cubical" decay (see later) and that extends more than 1 m up the stem is likely to be *H. annosum*. Other signs worth looking for are resinosis at the base of standing trees (though this is a general response to root diseases, not specific to *H. annosum*) and fruit bodies on stumps or trees, or on the ground (figs 21, 22 and 23). If severe infection by Fomes is suspected, confirmation and advice on future stand management should be sought from a specialist.

There are only two other causes of butt-rot in conifers that are of any significance: *Armillaria* and *Phaeolus schweinitzii*. The latter, which is most likely to be encountered in older Sitka spruce, larch or Douglas fir, causes a rot in which affected wood is dry and dark brown with horizontal and vertical fracture lines: a "brown cubical" rot. At an advanced stage, cubes of decayed wood can be crushed in the fingers to produce something like coffee powder. *Phaeolus* decay is a frequent cause of stem breakage in large conifers (fig 71). *Armillaria* is more difficult to recognize, but wood decayed by it is usually extremely wet – indeed, in some cases, water can be squeezed out of it as from a sponge. When wood decayed by *Armillaria* is broken longitudinally and radially (i.e. down the grain), prominent white flakes of fungal tissue a few millimetres across can usually be seen in the rays. In old decay columns there may be a cavity containing rhizomorphs and sheets of black fungal tissue. *Armillaria* decay rarely extends more than a metre up the stem in conifers and is probably not often a cause of serious commercial loss. However, if its presence is suspected in large trees, further investigation might be warranted because of its ability to decay roots and render trees liable to windthrow. The presence of *Armillaria* would also imply some risk of mortality in the early years of a following crop.

There is no single fungus like *H. annosum* that dominates practical consideration of decay in broadleaves. There are several common fungi that are capable of inducing severe root- and butt-rot in particular species – examples are *Ustulina deusta*, *Inonotus dryadeus*, *Ganoderma* species and *Meripilus giganteus*. Instances of serious decay by these or other species might be indicated by the production of fruit bodies on or near the stem base (fig 56) – though

Fig 71. Breakage of a larch stem caused by *Phaeolus schweinitzii* decay.

this should not be relied upon. Readers are referred to the sources listed in section 8.2 for descriptions of the fruit bodies of some of the more common decay-causing species.

8.4. *Phellinus (Fomes) pini*; "Ring-rot"

Some decay fungi are adapted to gain entry to the stem wood via branch stubs or dead branches. Although not generally an important class of decays in British forests, one species, *Phellinus pini*, is occasionally encountered in older conifers. It tends to enter stems through branch stubs or scars fairly high in the crown, though its association with them may not be as clear as that of a wound-rot with a machine or deer wound. Wood decayed by *P. pini* usually takes the shape of a ring or part-ring, or two or more concentric rings, in cross-section. Its fruit bodies, which may form on branch scars, are hard, greyish and hoof-shaped with pores on the under-surface.

Printed in the United Kingdom for the Stationery Office
J 31043, C20, 3/98